Career Planning for Research Bioscientists

Career Planning for Research Bioscientists

Sarah Blackford

Society for Experimental Biology
Bailrigg House
Lancaster University
Lancaster LA1 4YE

&

Charles Darwin House
12 Roger Street
London WC1N 2JU

WILEY-BLACKWELL

A John Wiley & Sons, Ltd., Publication

This edition first published 2013 © 2013 by Sarah Blackford

Blackwell Publishing was acquired by John Wiley & Sons in February 2007. Blackwell's publishing programme has been merged with Wiley's global Scientific, Technical and Medical business to form Wiley-Blackwell.

Registered Office
John Wiley & Sons, Ltd, The Atrium, Southern Gate, Chichester, West Sussex, PO19 8SQ, UK

Editorial Offices
9600 Garsington Road, Oxford, OX4 2DQ, UK
The Atrium, Southern Gate, Chichester, West Sussex, PO19 8SQ, UK
111 River Street, Hoboken, NJ 07030-5774, USA

For details of our global editorial offices, for customer services and for information about how to apply for permission to reuse the copyright material in this book please see our website at www.wiley.com/wiley-blackwell.

Library of Congress Cataloging-in-Publication Data

Blackford, Sarah.
Career planning for research bioscientists / Sarah Blackford.
    p.   cm.
  Includes bibliographical references and index.
  ISBN 978-1-4051-9670-3 (pbk.)
1. Biology–Vocational guidance.   2. Career development.   I. Title.
  QH314.B53 2013
  570.23–dc23

                                                                2012016866

A catalogue record for this book is available from the British Library.

Wiley also publishes its books in a variety of electronic formats. Some content that appears in print may not be available in electronic books.

Cover image: Tom Donald (http://clearwood.co.uk)
Cover design by Steve Thompson

Set in 10.5/13pt Janson by SPi Publisher Services, Pondicherry, India
Printed and bound in Malaysia by Vivar Printing Sdn Bhd

1   2013

Contents

COMPANION WEBSITE:
This book has a companion website:
www.wiley.com/go/blackford/careerplanning and you can visit
the author's blog at www.biosciencecareers.org for more information.

Author's note

Career planning is as vital to researchers as planning experiments. To produce successful results, you need to be familiar with the career landscape, know how to promote yourself and be skilled in your techniques. Competition for permanent academic posts is fierce and to be successful you must manage your career carefully to maximise your chances of success. As doctoral students and postdoctoral researchers, numerous career options are available to you but navigating your career path can be challenging. This book aims to assist you in this process whether you are planning a career within or outside academic research.

As a qualified careers adviser working in an international bioscience learned society, having previously worked in scientific research and publishing within university bioscience departments, I am well aware of the particular career issues faced by researchers. For my own part, my career track has been far from straightforward. Having realised, during my first 3-year contract, that research did not suit me (and, following a few mishaps in the lab, I did not suit it!), I changed careers into journal publishing, as I had enjoyed communicating my work. This editorial role was more suited to my skills and personality and was a convenient stepping stone out of the lab. However, my desire to help and support people was still not realised and it was only when, by chance, I saw an advert for a careers adviser post that I considered the prospect of changing career paths again. Although I didn't get this job, the insights I gained into the sector during the interview process acted as a catalyst for further action. Through volunteering, networking and then, later, formalising my qualifications with a Master's degree, I secured two successive careers advisory posts, ultimately arriving at my current role as the head of education and public affairs of the Society for Experimental Biology (SEB). My career path may seem rather disjointed but it is not uncommon for people to experience two or three different careers during their working life. I am using many of the skills I developed during my previous jobs and even publishing research papers derived from my careers work.

Since taking up my role at the SEB, I have been delivering career workshops for research bioscientists at career conferences, in universities and during international scientific meetings. In addition, I offer one-to-one career coaching and guidance to individuals. This book communicates this advice and knowledge in one complete compendium, which you can refer to at your convenience. I hope you enjoy reading it and find it a useful guide for your career. To keep it as accurate as possible (all information was up to date at the time of the book's submission in 2012), I welcome any suggestions (via my blog or Twitter - see following page) for further resources which you think would be helpful to readers of the book.

Acknowledgements

At the risk of reading like a Hollywood acceptance speech, I have many people I would like to thank who helped me turn this book from an idea into a reality.

For their invaluable input and contributions to this project, I would like to acknowledge the following researchers, academics, careers advisers and associated professionals: Dr Tennie Videler (Vitae), Dr Teresa Valencak (Veterinary Medicine University, Vienna), Dr Barbara Tigar (Liverpool Hope University), Dr Jenny Sneddon (Liverpool John Moores University), Dr Sara Shinton (Shinton Consulting), Professor Dale Sanders (Director, John Innes Centre), Dr Marie Reveiller (NYU Langone Medical Center), Dr David Proctor (independent contractor), Dr Jeremy Pritchard (University of Birmingham), Cathee Johnson Phillips (Executive Director, National Postdoctoral Association), Charlotte Lindsay (freelance editor), Dr Calum Leckie (University College London), Dr Alison Kingston-Smith (IBERS, University of Aberystwyth), Dr Hilary M. Jones (University of York), Sarah Jones (Association of the British Pharmaceutical Industry), Clare Jones (University of Nottingham), Dr Anne Forde (University of Cambridge), Professor Tony Farrell (University of British Columbia), Dr Gordon Dalton (University College Cork), Dr Simon Cutler (BBSRC), Dr Anne Borland (University of Newcastle) and Dr Alun Anderson (former Editor-in-Chief, *New Scientist*). Not forgetting also, all those whose Career Narrative is featured in Appendix 1.

I would especially like to acknowledge Dr Irene Hames (editorial consultant), who inspired me to start the book and gave me the benefit of her time and expertise during the writing process. My thanks also go to those who were involved in the production of the book: Holly Regan-Jones, Suvesh Subramanian, Kelvin Matthews, Carys Williams and Ward Cooper of Wiley Blackwell.

For their support and encouragement, I am indebted to my friends and family. I would also like to express my appreciation to the Society for Experimental Biology, with whom I have worked for over 10 years and who champion the professional development of bioscientists. In particular, my thanks go to Professor Bill Davies (Lancaster University), who initiated the SEB's education programme and has been a much-valued mentor during my career.

Sarah Blackford
www.biosciencecareers.org
@bioscicareer

Believe in yourself and anything is possible

CHAPTER 1

Introduction

Who is this book for?

Are you a research bioscientist looking for help with your career? Do you want to know where to find out about jobs? How to write an effective CV? Improve your interview technique? Or learn more about how to take control of your career? If so, this book should help you. It contains advice and information specifically tailored to the needs of research bioscientists. It offers you strategies to enhance your chances of success, following a recognised theoretical model for effective career planning. It contains information about research and non-research jobs, funding and courses, how to present yourself in a CV, enhance your employability and be successful at interview. Also included are less tangible, but highly important, aspects of the career planning process such as self-analysis, decision making and action planning.

The world of work has changed and will alter even more rapidly in the future. The concept of a 'job for life' has long gone in many professions in most countries, replaced by a relatively insecure employment culture. However, as highly qualified and skilled researchers, you have the opportunity to forge a successful career in many occupational fields. Whatever 'successful career' means to you, it will be your actions and your purposefulness which get you there. Don't rely on your supervisor or manager to organise things for you; you need to be proactive and plan your career strategy. This book aims to help you to succeed by providing you with the information, tools and resources which will assist you with your career planning.

Planning your career can be challenging as not all career choices are within your grasp. You may not know in which direction you want to take your career or how to do it. Career plans and career direction vary from one person to another. If you are considering an academic career path, it may not turn out to be possible. Competition for permanent academic research positions is notoriously harsh in many countries. Statistics show that only between 7% and 14% of postdoctoral researchers will achieve a full academic career and secure a tenured position (Bradley 2009; Kirshenbaum 2008; Newman 2007). The situation is worse for women (see Chapter 5, Box 5.5). Therefore, it is advisable to keep your options open and have more than one career plan. Maybe you are considering other career options, such as a job in industry, science administration, policy or communications work. Some of you may be unsure about what you want to do. Personal circumstances change throughout the course of your life; sometimes you will be

Career Planning for Research Bioscientists, First Edition. Sarah Blackford.
© 2013 Sarah Blackford. Published 2013 by Blackwell Publishing Ltd.

Box 1.1 Definition of terms used in this book

Research bioscientist

This term refers to researchers working across all subject disciplines within the field of the (primarily lab-based) biosciences and includes physiologists, microbiologists, molecular biologists, animal and plant scientists, biochemists, biomedical scientists, geneticists, ecologists and pharmacologists. It excludes vocational bioscience-related disciplines such as medicine and associated fields such as physiotherapy, pharmacy, veterinary science, dentistry and clinical careers as these can have significantly different career structures.

Doctoral student

For the majority of this book, I use 'doctoral student' to refer to those who are studying for a PhD (or D.Phil) by research. Equivalent terms more recognisable to some are postgraduate student, postgraduate researcher, doctoral candidate/scholar and PhD student. The programme of study is normally 3–6 years in length (but can be up to 8 years in some countries). PhDs are accredited by universities and are carried out there, in research institutes, industry, hospitals and government agencies.

Postdoctoral researcher

The term 'postdoctoral researcher' refers to a PhD-qualified researcher who is employed to conduct academic research. Equivalent terms more recognisable to some are early-career researcher, postdoctoral associate, postdoc, postdoctoral, research associate, postdoctoral scholar and contract research staff.

Career planning

Career planning is a life-long learning process which involves personal development, the ability to make informed career decisions, self-manage and ultimately ensure your employability. Career planning and associated terms, such as career management and career development, are used interchangeably in this book.

Employability

Employability does not mean 'employment'. There is no single definition but here it is defined as a set of personal aptitudes, skills and abilities which will enhance your chances of securing and maintaining employment.

---

fully flexible and able to relocate, while at other times personal considerations may compromise your career decisions.

As research bioscientists you have things in common which make this book relevant to you (see Box 1.1 for a definition of terms).

- Research bioscientists wishing to pursue an academic career are exposed to a highly competitive market. This book suggests ways in which to maximise your chances of success.
- Leaving academic research, although a common activity amongst research bioscientists, can be a tough decision to make. This book showcases a range of career options for you to consider as well as providing further resources to investigate.
- Many researchers define themselves by their subject discipline or research project. But who are 'you'? Self-knowledge is integral to finding a career which suits you and this book will help you to recognise your strengths and areas you need to develop.

- Once you are more informed about the job market and your skills and abilities, strategies are put forward to enable you to harness new opportunities.

It would be an impossible task to write a careers book with advice and information tailored specifically to every person's situation (that is the preserve of the individual careers interview). Equally, the entry requirements for different professions can vary at the international level, so what is true in one country does not necessarily hold for another. Research is an international endeavour with researchers working in multinational groups all over the world. My own careers work has been centred primarily in western Europe, and the UK in particular. However, work cultures in the majority of countries in the developed world are sufficiently similar nowadays that the content of this book will be relevant to all research bioscientists seeking advice with their career planning.

The process of career planning

There is more to career planning (see Box 1.1 for a definition of terms) than looking for jobs and making applications. Many people are not clear about what they want to do next. Those who have a specific career goal in mind need to think carefully about how to achieve their aim. Even when you know what you want to do, circumstances can intervene to change your career direction. The process of career planning involves taking control and managing your career so that you are making the most of your current post in preparation for your next career move or 'transition'. This means knowing when to take advantage of opportunities, being strategic and proactive, making informed decisions and being aware of your own particular strengths and weaknesses. Consider some of the following questions.

- How well are you managing your career?
- What are your career plans for the near and more distant future?
- Looking back over the past 3 years, what experiences, skills or knowledge have you acquired that has contributed to your personal growth?
- How happy are you with the progress you have made over these years? Could you have done more?
- Are there gaps in your experience which could hinder your career progression?

Helping you with your career planning

In my role as a careers adviser working on behalf of a bioscience learned society, I have developed a repertoire of career development workshops aimed at bioscience doctoral students and postdoctoral researchers, which I deliver in universities, independently and during scientific conferences. Similarly, other organisations, policy and funding bodies do the same, and many universities and research institutes employ staff and offer career programmes designed for their research staff and students. The existence and visibility of these development programmes vary across university departments, institutions, countries and continents. In Europe, the European Charter for Researchers (2005) and The Concordat (2006) provide guidance on good practice for higher education institutions, research institutes and other organisations employing postdoctoral researchers. In the US, the Carnegie Initiative on the Doctorate (CID) worked with doctoral training departments to restructure their programmes to better prepare doctoral candidates after the National Academy of Sciences, National Academy of Engineering and Institute of Medicine concluded that more funding should be made available to support the career development of postdoctoral researchers (Golde & Walker 2006; Walker et al. 2008). Furthermore, bodies such as the European Universities Association Council for Doctoral Education (EUA-CDE 2012) advise on the continuing improvement and development of doctoral education and research training programmes.

The aim of this book is to transmit the information, guidance and advice from these career development programmes in order to provide a useful and convenient compendium to refer to. This book discusses the many complex factors which exist within the career planning process, enabling you to find effective strategies and make considered decisions for a successful career.

Content of the book

Chapter 1: Introduction

This chapter establishes the aims and objectives for the book and its content. The importance of paying attention to your career is discussed in the context of initiatives already in place to assist doctoral students and postdoctoral researchers to capitalise on their personal and professional assets.

Chapter 2: Planning your career

Theories of career development and planning underpin the information and guidance in this book. This chapter focuses on two key career planning models and establishes a structure for the book.

Chapter 3: Self-awareness

How well do you know yourself beyond your research interests and technical skills? How will you decide which careers are suited to you? This chapter contains exercises to help you examine self-awareness dimensions, such as your skills, personality and values, so you can make informed decisions about your career and write effective job applications.

Chapter 4: The job market

Where are the jobs and how do you engage effectively with the job market? With so much competition, how do you stay positive in your job search? Which careers should you consider? This chapter analyses a range of jobs and looks at how you can discover the 'hidden' job market through networking.

Chapter 5: Enhancing your employability

In Chapter 5, suggestions for personal and professional development to enhance your employability are put forward, including ways to extend your experiences beyond your core research work.

Chapter 6: Making applications

There are a number of ways in which to apply for a job. The most commonly used are the application form, curriculum vitae (CV) and resumé. Usually sent via email or an electronic submission system, it is vital that the documents adhere to guidelines and present your information convincingly. This chapter gives advice about how to make effective applications and is supplemented with example CVs.

Chapter 7: Successful interview technique

This chapter provides advice on techniques and strategies you can use before, during and after your interview to achieve a successful outcome. Sample questions give you the opportunity to consider how you would answer them effectively.

Chapter 8: Decision making and action planning

Based on a recognised coaching model, this chapter brings together the information gathered from previous chapters (especially Chapters 3 and 4) to enable you to make informed decisions about your career.

Appendix 1: Career narratives

Appendix 1 comprises 20 career narratives from research bioscientists working in a range of professions. Each case study describes the person's job, how they successfully moved into this profession and includes a commentary on their career strategies.

Appendix 2: Social media

Chapters 4 and 5 are supplemented by Appendix 2, which describes how researchers can make use of social media to enhance their employment prospects and access the 'hidden' job market through networking.

Appendix 3: Example CVs

Appendix 3 supplements Chapter 6 and provides six example CVs, demonstrating how they have been adjusted to six different job advertisements and their corresponding specifications.

Appendix 4: Support and resources

Appendix 4 consists of a comprehensive list of support groups, web resources and a bibliography for further information.

How to use this book

Much of the information in this book is generic and is a useful reference for anyone looking for an effective career planning strategy. For example, changing patterns of work as a result of a global economy, models of career planning and the capacity for self-reliance apply to everyone. However, as research bioscientists, you possess a particular set of skills and experiences which have been used in the book to illustrate how you can capitalise on them. Example CVs and career profiles are based on research bioscientists working in a range of professions. Job vacancies, sources of support and further information are primarily bioscience related.

The way you use this book is up to you. You can read it from cover to cover for a full overview of how to plan and manage your career. You may prefer to dip into particular chapters, or perhaps you are searching for specific information such as how to write an effective CV or improve your interview technique. However you make use of it, this book aims to provide you with concepts and information, practicalities and tools to assist you in performing one of the most important experiments of your life – your career.

References

Bradley M. (2009) Contractual obligations. *Sciencebase*. Available from: www.sciencebase. com/contract_research_assistant.html.

European Charter for Researchers. (2005) Available from: http://ec.europa.eu/eracareers/ pdf/am509774CEE_EN_E4.pdf.

European Universities Association Council for Doctoral Education (EUA-CDE). (2012) Available from: www.eua.be/cde/.

Golde C, Walker G. (2006) *Envisioning the Future of Doctoral Education: preparing stewards of the discipline – Carnegie essays on the doctorate*. San Francisco: Jossey-Bass.

Kirshenbaum S. (2008) Plight of the postdoc. *Science Progress*. Available from: http:// scienceprogress.org/2008/06/plight-of-the-postdoc/.

Newman M. (2007) Postdocs embittered by lack of career prospects. *Times Higher Education Supplement*. Available from: http://bit.ly/ta9Qel.

The Concordat. (2006) *The Concordat: to support the career development of researchers*. Available from: www.vitae.ac.uk/concordat.

Walker GE, Golde CM, Jones L, Bueschel AC, Hutchings P. (2008) *The Formation of Scholars. Rethinking doctoral education for the twenty-first century*. San Francisco: Jossey-Bass.

CHAPTER 2

Planning your career

The importance of career planning

Much of a researcher's time is spent planning and managing. You carry out a whole range of tasks, such as designing and conducting experiments, analysing data, problem solving, multi-tasking, supervising others and reading research literature. Activities such as communicating your research, networking, self-development and self-motivation all work together to extend your skill-set and can ultimately determine your career direction. If you apply these broad skills to the wider areas of your life you will, effectively, be managing your career.

Within industry, especially in many large organisations, career progression and personal development are addressed in a formal manner, tied into appraisals at yearly or biannual intervals. During these meetings, usually with a line manager, employees review their performance and identify areas for development. For example, in the short term they may wish to attend a course to learn a new technique. In the longer term, the overall direction of their career may be the subject of discussion. Objectives are set, to be reviewed at the next meeting, and so the process continues. More discerning companies invest in training for their employees, so they are responsive to new innovations in a fast-paced global market.

Increasingly, universities and research institutions are being exposed to a similar set of pressures, including global competition and economic demands, causing them to align more closely with the commercial sector. Formal appraisal schemes have been more rigorously implemented, extending to postdoctoral research staff and, in some cases, doctoral students, although a recent survey (ASSET 2010) conducted in the UK showed that only 20% of postdoctoral researchers who responded had received an appraisal.

In many countries, career development programmes for postdoctoral researchers and doctoral students have been put into place in graduate schools or across departments and institutions. These development programmes consist of a wide range of activities to help increase self-awareness, enhance self-efficacy and assist career advancement. Training such as thesis writing, presentation skills, publishing in journals, fund raising and science communication help to enhance academic-related skills. More generic training in effective networking, leadership and self-analysis can provide the means to develop careers at a broader level. All of these skills are transferable to a wide range of careers, so take advantage of career development opportunities available to you.

Career Planning for Research Bioscientists, First Edition. Sarah Blackford.
© 2013 Sarah Blackford. Published 2013 by Blackwell Publishing Ltd.

Many researchers say they do not receive careers support but in many cases, they are not aware of what is on offer. Even if your own institution does not provide support and training, external organisations such as learned societies, specialist organisations, funding bodies and private companies offer a wide range of career development opportunities specially designed for researchers. These may include free online resources, bursaries to support attendance at conferences or career workshops and events (see Chapter 5 on personal and professional development).

Your personal career development will depend on the direction in which you want to take your career. For example, those planning for an academic career need to develop the skills and knowledge associated with senior academic positions, such as writing papers (and getting them published), applying for and securing funding, developing an international profile and taking on teaching responsibilities. For other careers, your skills may be more usefully developed in areas such as writing for general audiences, demonstrating science in schools or spending a portion of your research time in a company. Whichever skills you possess, and are developing during the course of your research, they will all demonstrate valuable evidence of your abilities and will be transferable to a wide range of career sectors.

What is career planning?

Career planning and education have been the focus of research since the early 1900s and new models are still emerging according to the changing world of work and our social environment. This book is organised according to two distinct career theories:

- the DOTS model (Law & Watts 1977) and
- planned happenstance (Mitchell et al. 1999).

These two models are relatively contrasting in their theoretical basis. The *DOTS model* (Fig. 2.1) is structured and logical and identifies four fundamental factors which should be considered when planning your career:

- **D**ecision making (weighing up options and deciding which to act upon)
- **O**pportunities (finding out about jobs, networking and creating opportunities)
- **T**ransition (making applications, going for interview)
- **S**elf (awareness of your skills, interests, personal attributes and values).

Although developed further (Law 1999), the fundamental theory of the DOTS model remains one of the most common frameworks upon which careers education curricula and planning programmes are based.

Planned happenstance is more random and intuitive and recognises the need to respond to chance events outside our control. The model includes the key factors which enable us to identify opportunities and take advantage of chance happenings:

- curiosity (exploring new learning opportunities)
- persistence (exerting effort despite setbacks)
- flexibility (changing attitudes and circumstances)
- optimism (viewing new opportunities as possible and attainable)
- risk taking (taking risks in the face of uncertain outcomes).

Fig. 2.1 Diagrammatic representation of the career planning DOTS model proposed by Law and Watts (1977). The arrows imply a sequential process but many or all of these actions will be taking place simultaneously.

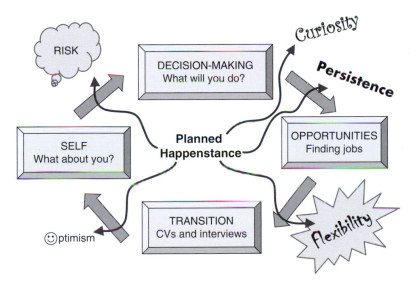

Fig. 2.2 Diagrammatic representation of the DOTS model (Fig. 2.1) blended with planned happenstance theory (Mitchell et al. 1999), illustrating the need for a combination of structure and flexibility in the career planning process.

In certain respects, these two theoretical models reflect the way in which you plan your research. There is the predictable structured research planning, whereby you follow set protocols to achieve your milestones and generate results. However, you need to be curious and flexible when unexpected findings lead you off in new and unpredicted directions, or persistent when experiments are not producing results (Fig. 2.2).

Career planning is much the same. If you know where you want your career to be in, say, 5 years' time, you can focus on specific activities identified in the DOTS model which will enable you to achieve that goal. However, not only do many of us not have a specific career goal in mind but even if we do, we are still exposed to random external events which we can tap into and use to our advantage. Recognising and harnessing these opportunities is central to the planned happenstance theory which shows you ways to manage these chance events.

Taken together, these two models underpin the information and guidance in this book to enable you to exploit the planned and unplanned aspects of your career. Examples where structured and flexible planning has figured in people's careers are related in career narratives 10, 11, 12 and 18 in Appendix 1.

Career planning in action

How do you approach your career planning? What do you need to do? In much the same way that keeping abreast of the literature in your research field will help to keep you in touch with your subject, similarly awareness of the employment market will keep you up to date with jobs being offered by employers. Being focused, purposeful and informed about each post you apply for will give you an advantage and indicate your suitability for it. You should be able to demonstrate that you have the experience, skills and qualifications required for the post. For example, knowledge of the job market, emerging technologies or research trends may help you to focus your applications or target particular companies with speculative enquiries. You could even create your own job through proactive self-marketing and networking.

Ideally, you should keep the momentum of your career moving and avoid stagnating passively in the same post for the majority of your early career. Too much engagement with the same skills, the same experience and the same environment could jeopardise your career prospects later on. Personal development planning (PDP) tools exist which can help you chart your experiences and identify areas for personal development. Specialised PDP tools have been designed for researchers for this purpose (see Chapter 3). These tools can be used individually or linked to appraisals.

Conclusion

Life itself is too complex and unstable to allow us to make fixed career plans. How many of us knew we would end up where we are now, in our careers or personal lives? Are we any better or worse off than we would have been had we taken a different path? We will never know. What we can be sure about is that we have some power over our next move forward. This is under our own direction but also in the hands of outside influences beyond our control. Such influences can be brief and localised, e.g. other people's actions, whether or not an experiment yields results or a chance encounter at a conference. Or they may be more overarching and powerful, such as the political and social state of a country, economic fluctuations and even sudden and unexpected world events.

The aim of this book is to help you plan for a sustainable career path, which you will find fulfilling. You have already embarked on your career journey and, perhaps, you are enjoying your current job immensely. You may love your research topic so much you want to pursue it for the rest of your working life. If academic research is your career goal, you may have to adapt or change your research interests in order to remain in contention for funding streams as they evolve over the years (see career narratives 1 and 3, Appendix 1). Maybe you are interested in other careers but you're not really sure what's out there and what exactly you'd be suited to do. You may be passionate about your subject but the stresses of academia are not for you. Everyone reading this book will either be a doctoral student or postdoctoral research bioscientist (or will have an interest in how research bio-scientists can plan their careers). That is all you have in common with each other. Everything else – your personality, skills, social and cultural background, values and interests – will be varied and different. Therefore, I do not intend to give you specific answers or instructions in this book, otherwise I would be including some people and excluding others. What I aim to do is to offer guidance, information and a framework which will provide strategies, insights and even moments of enlightenment to support you in planning your next career move.

References

ASSET. (2012) *The Athena Survey of Science Engineering and Technology*. Available from: www.athenasurvey.org.uk/uk_results_2010.pdf.

Law B. (1999) Career-learning space: new-dots thinking for careers education. *British Journal of Guidance & Counselling* 27(1), 35–54.

Law B, Watts AG. (1977) *Schools, Careers and Community*. London: Church Information Office.

Mitchell KE, Lewin AS, Krumboltz JB. (1999) Planned happenstance. Constructing unexpected career opportunities. *Journal of Counselling and Development* 17(2), 115–24. John Wiley & Sons, Inc.

CHAPTER 3

Self-awareness

Self-awareness is the key to career success and fulfilment so if you only have time to read one chapter of this book, read this one. Self-awareness features in most career planning and management models, including the DOTS model described in Chapter 2. However, it is usually neglected as people take more practical action such as scanning job adverts and making applications. The latter activities are, of course, an important part of career planning but it is self-awareness which is fundamental to the process. It underpins your ability to make informed decisions about your career, enables you to assess which jobs will suit you, how you prefer to conduct your work and where your strengths lie. Self-awareness also offers clues as to why these are your preferences, which may further assist you in making decisions about your personal development and career choices. It can even improve your performance at interview.

What is self-awareness?

There are many definitions of self-awareness. In the context of career planning, it is defined by a range of personal factors such as interests, skills, personal qualities, values, personality, intelligence, personal situation, gender, cultural, moral and social background. These can, in turn, be broken down into subcategories as shown in Figure 3.1, which illustrates four of the main self-awareness factors to be considered in career planning (Watts 1977).

Knowledge and interests

The two factors with which most of us define ourselves (especially in work) are knowledge and interests. They are also perhaps the most influential in terms of career choice. Your interest in and knowledge of the life sciences and the particular specialisation you have chosen is what has, most probably, led you into your career as a research bioscientist. In particular, within academic research, this interest is highly focused and concentrated, requiring intensive commitment. However, basing your career solely on your knowledge and disciplinary interests can limit your options. Deep career thinking requires a broader awareness of 'you' so that you widen your career perspectives and broaden your options. Research tasks and other activities may more accurately reveal where your career interests lie: you may enjoy planning and designing experiments over practical lab work; perhaps problem solving and analysis are what truly interest you or maybe communicating science is your real passion. Your personal time (if you have any!) may be spent

Career Planning for Research Bioscientists, First Edition. Sarah Blackford.
© 2013 Sarah Blackford. Published 2013 by Blackwell Publishing Ltd.

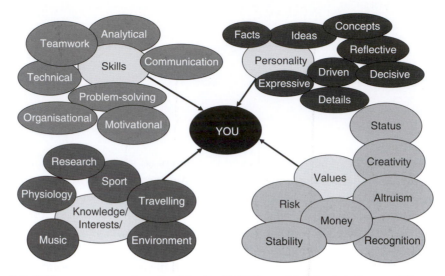

Fig. 3.1 Self-awareness as described in the DOTS model (Fig. 2.1; Law & Watts 1977) comprises four main categories which can be further dissected into their subcomponents, a selection of which is shown here.

pursuing extracurricular interests such as sport, photography, music, travelling and socialising. These activities also offer clues about your career choices, revealing your personal preference for leadership, creativity, autonomy, management or team working. Sometimes people even turn their personal interest into a career (see career narrative 20 in Appendix 1).

Skills

During the course of your career as a research bioscientist, you have been learning techniques and methodologies associated with your research project(s), and also developing other more generic research-related skills, e.g. communication, organisation, time management, resource management and networking. Employers are very keen on these skills and may also specify personal qualities such as self-motivation, creativity and tenacity alongside them. Job descriptions are packed full of skills which vary according to the type of job being advertised (see Chapter 4). Specific laboratory skills are essential for technical or research posts, which may also require particular knowledge of the research field. For more senior academic posts, evidence of grant writing and publishing, collaborative and teaching experience will be specified. For other career options, more emphasis will be placed on generic skills, such as communication, organisation, problem solving, analytical and management skills.

Being able to recognise the skills you possess and, in particular, those you enjoy using or would like to develop further can provide clues to the type of jobs and roles to which you are most suited and help you to make effective career choices.

Personality

Awareness of your personality type can increase your personal effectiveness in many areas of your life, including career planning. As well as helping you to make informed career decisions, it can have a more subtle influence such as improving your understanding of others and thereby your personal and professional relationships. Being aware of your personality type and your preferred way of doing things can help you to organise your time and workload more effectively and reveal areas for personal development.

Values

Schwartz (1992) defined personal values as 'desirable, trans-situational goals which vary in their importance amongst individuals as guiding principles in their lives'. Schwartz derived 10 types of values: power, achievement, hedonism, stimulation, self-direction, universalism, benevolence, tradition, conformity and security. Your motivational values are at the heart of your 'self' and will influence decisions you make about the kind of work which attracts you, your preferred professional role and working environment (e.g. commercial, charitable or educational). Values can alter during the course of your lifetime depending on your situation.

Other contributing factors

Your personal circumstances change during the course of your life. According to Super (1981), the different stages of your life influence the career decisions you make. You may experience periods during which you need more stability (e.g. raising a family), or you may need to make a compromise and revise your work–life balance due to a partner's career. At other times, you might be more mobile, able and willing to destabilise your personal life and relocate to another institution or country in order to further your career. Ideally, you will be more mobile at the start of your career in order to extend your experience and expand your skills so that you are in a better position to achieve a permanent and sustainable position later on (although academic leadership positions or a major career change may still require a substantial upheaval in later life). Furthermore, factors such as gender, social and cultural background are integral to your identity and will play a fundamental role in decisions you take regarding your career. Career narratives 6, 9, 10 and 18 in Appendix 1 demonstrate where people have guided their careers according to personal circumstances. Organisations and initiatives exist to help people keep their careers on track, for example those aiming to come back to work after a career break (see Appendix 4: Support and resources).

Practical ways to analyse your 'self' and increase self-awareness

We all carry out self-analysis routinely through self-reflection or by talking things over informally with friends and colleagues. We may wonder to ourselves why something happened, what we thought or felt about it. Professional coaching and counselling are available for more formal self-analysis, which may also take place during an appraisal with a line manager or supervisor. Mentoring and coaching schemes exist in some businesses and educational institutions or can be accessed online (e.g. MentoNet.net). These schemes have proved very helpful to researchers in supporting career development and increasing self-confidence (Vitae 2011a).

Carried out individually, a personal audit helps you to reflect on and record your activities. Reflection of this kind is promoted at many educational institutions where staff and students are encouraged to keep a 'record of progress' or 'personal development plan'. The National Postdoctoral Association (NPA) offers support to doctoral students and postdoctoral researchers through its 'Core Competencies'. Similarly, Vitae (2011b) has developed the Researcher Development Framework. Both tools can be accessed online. Specifically designed for researchers, they enable you to record your progress during the course of your career. More generic personal audit tools are also available such as those described by Bolles (2012) and Hawkins (1999). These are aimed at a more general audience but researchers may still find the exercises useful and enlightening. Similarly, Prospects Planner asks you questions about your skills and attributes to help you to generate new career ideas. See the references listed at the end of this chapter to access these self-assessment tools.

Individual personal reflection of this kind can be perceived by some people as a tedious or unnecessary task. However, reviewing your current and previous activities, the skills you have developed and the decisions you have made up until now can reveal clues about your preferred career path and help you with current and future decisions. Keeping your personal audit up to date can also assist you when making applications and writing your CV, enabling you to recall your skills and achievements more accurately.

As well as individual reflection via discussion or a personal audit, psychometric tools and other personality-related instruments help you increase your self-awareness. Many people find doing these kinds of exercises easier than independent self-analysis. Psychometric tools normally employ questionnaires to gather information about various aspects of a person's character and then translate this into a personal profile to assist in the process of enhancing self-awareness.

In this chapter you can try out three different self-analysis exercises:

1. skills audit
2. personality assessment (summarised version)
3. values analysis (abridged version).

1. Skills audit

Reviewing the job market by examining advertisements is a useful way to find out what skills and personal qualities employers are looking for (see Chapter 4). Knowledge of the job market and your own skill-set gives you a valuable insight into how well you match up to, and your chances of securing, a particular job. For example, if you are looking for a postdoctoral position, your research record, specialist technical skills, knowledge and qualifications will be the most important requirements and deemed essential to the job. Research posts in industry require a broader set of skills, and the ability to work in a team will be essential. For non-research careers, your research expertise becomes less relevant and needs to be translated into transferable skills which take centre stage.

Figure 3.2 illustrates how specialist knowledge and skills give way to generic non-specialist skills the further you move away from academic research. Of course, many non-bioscience jobs require their own professional specialist knowledge and skills, which may mean you having to upskill or take a course in order to move to a different career path.

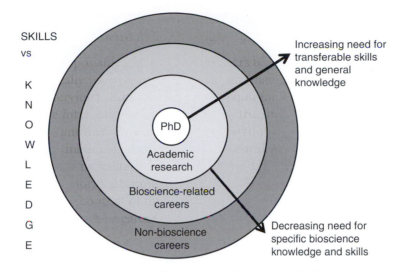

Fig. 3.2 The relative importance of specialist and non-specialist bioscience research skills and knowledge changes as you move away from an academic research career.

Being able to recognise the skills you possess, and identifying those you enjoy using and would like to develop, should help you to focus on relevant careers and areas for your personal development.

Skills task

I use this task during career workshops on self-awareness with doctoral students and postdoctoral researchers. The participants sit together in groups of 4–5 according to their career stage, i.e. doctoral students, junior postdocs, mid-postdocs, senior postdocs, etc. They work as a team and collectively identify all the individual tasks they perform in the course of their research. They write each task down on individual sticky notes which they then arrange into skill categories such as 'Research', 'Management', 'Planning', 'Organisation', 'Communication' and so on. Figure 3.3 illustrates an example from one workshop. The aim of this task is to reveal the vast range of skills you possess as researchers and it is a great opportunity to share experiences. It can also illustrate how skills develop as you progress from doctoral student level up to senior postdoctoral researcher when management-related tasks such as supervising, teaching, applying for funding and reviewing papers are more commonly carried out.

You can carry out this task individually or with a colleague or friend, using sticky notes to brainstorm everything you do. Next, arrange the individual tasks into labelled skill-sets as shown in Figure 3.3. Refer to Box 3.1, which lists some of the top skills and personal qualities employers seek, and add in your own skill categories too.

Another way to carry out your skills audit is illustrated in Table 3.1. Examples of tasks and their related skills are shown and, again, you can refer to the list in Box 3.1 as well as adding in others of your own choosing.

The final part of this exercise is to rate your enjoyment of each activity and relate this back to the associated skills by using a scoring system (see Table 3.1). The scoring ranges between the lowest rating, 1, denoting you don't enjoy it at all, through to the highest rating, 5, denoting it is something you enjoy very much. Your enjoyment of the tasks you do (i.e. your job satisfaction) will give you clues about the direction of your career. We all have duties within our jobs

Fig. 3.3 Results of a skills audit during a workshop where doctoral students and postdoctoral researchers brainstormed all the tasks they carry out in their research and grouped them into corresponding skill-sets.

which we find less appealing (even if we are good at them). Ideally your next post will consist of tasks which employ more of the kinds of skills you find rewarding and fulfilling.

By conducting an audit of the tasks you perform in your current role, previous posts and outside activities, you will start to gain a picture of your personal skill-set. You can also relate these skills to job descriptions and make a more accurate assessment of the types of jobs to which you might be most suited. A useful resource developed by the Science Council describes '10 types of scientist': explorer, investigator, developer, communicator, teacher, service provider, monitor, entrepreneur, business and policy maker (www.sciencecouncil.org/10-types-scientist). Reviewing the tasks and skills which give you job satisfaction could help you to identify future roles to which you are well matched.

Box 3.1 Top personal skills sought by employers

The following personal skills and qualities are some of the most commonly sought after by employers across all employment sectors.

Communication skills (including speaking, writing, listening)

Communication is the main skill which you will see when you peruse the job advertisements across all sectors. Whether it is being able to make presentations, negotiate, produce reports, interact effectively with others within and outside an organisation, communication is at the heart of most professional roles.

Flexibility/adaptability

In these days of global change and new innovations, it is essential that organisations are flexible and adaptable and this is reflected in their workforce, which also needs to be open to new ideas and concepts. Employees also need to be able to change projects, multi-task and be able to work effectively in a team and independently.

Analytical/research skills

An ability to assess a situation, seek multiple perspectives, gather more information if necessary and identify key issues that need to be addressed is a key skill which you will be able to demonstrate through your research experience.

Self-motivation

You sometimes see advertisements asking for 'self-starters', i.e. self-motivated people who do not need constant supervision. Most professional posts will require self-motivated, energised people who will take responsibility for the success of projects.

Teamworking

More important in industry but becoming increasingly so in academia, the ability to operate effectively with other people in a professional manner, within one or more work groups, is usually fundamental to achieving a common goal.

Planning/organising

This relates to your ability to design, plan, organise and implement projects and tasks within an allotted timeframe, including goal setting.

Commercial/entrepreneurship

This skill is one of those mentioned by industry as being weakest amongst academic researchers (CIHE 2010). Your limited experience may still be able to demonstrate an understanding of funding and enterprise through grants you may have applied for (e.g. travel), budgeting for resources and links with industrial partners. Finding enterprising solutions to situations within your own research project can demonstrate an enterprising and intuitive mind.

Leadership/management skills

These skills relate to an ability to take charge and manage your co-workers effectively. You may have limited experience of leadership but you will most certainly have been managing your research project, possibly supervising students or instructing technical staff.

Self-confidence/self-belief

If you don't believe in yourself, why should an employer? Even more important if you are attempting to move to a new career path, it is necessary to have confidence in your abilities. If you have faith in yourself and what you can offer employers, they will have faith in you.

Table 3.1 Personal audit linking tasks to skills and level of personal enjoyment. Completion of this task is designed to facilitate awareness of skills and to identify skill gaps or areas for personal development in terms of preferred career path.

Activity	Skills (and personal qualities)	Enjoyment rating (1–5)
Designing experiments	Organisational, planning, negotiating for resources and technical help, self-motivation, creativity	
Biochemical and molecular techniques	Technical skills and methodologies, specialist knowledge	
Ordering chemicals	Managing resources, planning	
Analysing data	Analytical, problem solving, numeracy, logical thinking	
Supervising students	People management, communication, leadership/mentoring	
Presenting at a conference/seminar	Communication, resilience, self-confidence	
Running a research programme	Leadership, management, multi-tasking, taking responsibility	
Collaborating with research partners/industry	Communication, networking, entrepreneurship	
Helping and assisting colleagues	Problem solving, analytical, creativity, empathy	
Reviewing journal papers	Analytical, decision making, demonstrating subject knowledge and expertise	
Writing journal papers	Communication, knowledge of the field, perseverance	
Giving talks in schools	Communication, organisational, resilience, self-motivation, creativity	
Mountaineering	Teamwork, tenacity, motivation, strength of character	
Photography	Visual creativity, technical skills	

Note: Personal qualities are included alongside skills in Table 3.1 as they figure in job vacancies and can sometimes get confused. To clarify, personal qualities are innate attributes such as honesty, patience, initiative and resilience. Although relevant to the type of work which might suit you, they cannot be learned and developed in the same way as skills.

2. Personality assessment

Awareness and understanding of your personality are a more fundamental part of your 'self' and less easy to unpick than skills. Most of us would probably say that we know our own personality. However, being consciously aware of it and being able to analyse and articulate it are another matter. You will already be aware of certain aspects of your personality and how they influence your behaviour. For example, are you more of a talker than a listener in conversations? Do you prefer to judge situations objectively or are you more empathetic, putting yourself into the other person's shoes? Do you tend to plan well ahead or are you more inclined to leave things until the last minute?

Your answers to these questions will depend on different circumstances, so it is more difficult to unravel fundamental personality traits using an audit than with skills. Personality profiling tools, e.g. Myers Briggs Type Indicator® (MBTI®), Belbin® and 16PF®, have been developed which help to clarify individual personality preferences and aptitudes, usually using a questionnaire. It is important that you only undertake properly tried and tested profiling tools, whose validity and reliability have been scrutinised. While all of them have their critics, as long as they are conducted in a proper and ethical manner you should benefit from receiving a profile which will help to enhance your self-awareness.

Myers Briggs Type Indicator® (MBTI®)

As a qualified MBTI® practitioner, I have chosen to use this particular personality tool here to enable you to make an approximate assessment of your personality. The MBTI®, based on the fundamental psychological theory originally proposed by Jung (1923), was harnessed and developed by Myers and Briggs-Myers who devised and refined a set of (now 88) questions. This determined four personality preference dichotomies which, together, combine to make 16 personality types (Box 3.2).

Here is a brief description of the four pairs of personality preferences.

1 **(E) EXTRAVERSION** or **INTROVERSION (I)**: where you derive your energy.
2 **(S) SENSING** or **INTUITION (N)**: how you take in information.
3 **(T) THINKING** or **FEELING (F)**: how you make decisions.
4 **(J) JUDGING** or **PERCEIVING (P)**: how you like to order your life.

What is your MBTI® type? The MBTI® has been employed by professionals for over 60 years and, translated into many languages, it is an internationally recognised personality assessment instrument. It can only be administered using an official online or hardcopy questionnaire, so it is not possible for you to take the full test here. If you would like to take the full test, make sure you find a reputable organisation or qualified practitioner who is on the official register of recognised professionals. However, a shortened version based on Jung typology is available online at www.humanmetrics.com, or you can make an approximate assessment using the descriptors below.

As mentioned at the start of this section, personality is more difficult to pinpoint than skills, since behaviours alter depending on the situation you find yourself in. Bearing this in mind, look at the descriptors below in Box 3.3 and decide which, on balance, describes you best, e.g. E more than I, N more than S, T more than F, P more than J (or vice versa). You don't have to be completely on one side or the other, and for one or two of the preferences you may not be able to decide at all – remember this is just a taster of the MBTI®. Combine these preferences together to form the whole type with four letters, for example, INTP, ESTJ, etc. The combination of preferences gives rise to a total of 16 personality types (see Box 3.3). Look at Box 3.2 to see the dominant characteristics associated with each personality type.

Box 3.2 Sixteen personality preferences derived from the MBTI® illustrating the dominant characteristics of each personality type

ISTJ **Systematic** practical, sensible, logical, analytical, detached	ISFJ **Sympathetic** practical, realistic, thoughtful of others, sensitive, kind	INFJ **Insightful** creative, symbolic, compassionate, co-operative	INTJ **Visionary** creative, conceptual, rational, concise, objectively critical
ISTP **Pragmatic** detached, objective, analytical problem solvers	ISFP **Considerate** trusting, realistic, considerate, kind, observant, factual	INFP **Idealistic** sensitive, concerned, caring, curious, long-range vision	INTP **Logical** analytical, mentally quick, insightful, ingenious, ideas
ESTP **Action orientated** practical, realistic, straightforward, assertive, rational	ESFP **Friendly** practical, realistic, generous, warm, optimistic, tactful	ENFP **Enthusiastic** curious, imaginative, friendly, caring, spontaneous	ENTP **Innovative** creative, curious, imaginative, questioning
ESTJ **Decisive** realistic, clear, assertive, systematic, matter-of-fact	ESFJ **Helpful** warm, personable, sympathetic, down-to-earth	ENFJ **Appreciative** warm, loyal, compassionate, variety, challenges	ENTJ **Enterprising** analytical, logical, innovative, global thinkers, theorisers

It is important to keep in mind that there is no right or wrong, good or bad personality type and the MBTI® does not aim to pigeonhole people into particular boxes.

Understanding personality type allows you to objectify and understand your own behaviour as well as that of others. For example, if you have a 'J' preference (see Box 3.3) and are a person who likes to work steadily towards a deadline, you may be frustrated by 'P' colleagues who tend to work more erratically in bursts of energy, meeting deadlines at the last minute. If you look around your own working environment you can, perhaps, identify academics who have a 'closed-door policy', preferring to work alone (introverted, I), and those who regularly come into the lab to discuss the research project more openly and freely

**Box 3.3 Basic descriptors of the four primary MBTI preferences
(E:I, S:N, T:F and J:P)**

EXTRAVERSION (E)	**OR**	**INTROVERSION (I)**
EXTERNAL ENERGY		INTERNAL ENERGY
Talk more than listen		Listen more than talk
Like to talk things through		Prefer to think things through
Generally sociable		Generally reserved
Need people more than privacy		Need privacy more than people
Prefer a breadth of interaction		Prefer to talk on a one-to-one basis
Expressive		Quiet
SENSING (S)	**OR**	**INTUITION (N)**
LITERAL		LATERAL
Details		Patterns
Facts		Concepts
Practical		Imaginative
Sequential		Random
Directions		Hunches
Stick to the point		Go off the point
THINKING (T)	**OR**	**FEELING (F)**
THINK WITH HEAD		FEEL WITH HEART
Objective		Subjective
Want justice		Want harmony
Cool analysis		Caring empathy
Impersonal		Personal
Critique		Appreciative
Analytical		Empathise
JUDGING (J)	**OR**	**PERCEIVING (P)**
FILING SYSTEM		PILING SYSTEM
Organised		Flexible
Like structure		Go with the flow
Decisive		Procrastinate
Like closure		Keep things open
Plan ahead		Last-minute

(extraverted, E). Extraverted people tend to speak more than listen and may dominate meetings, while those with a preference for introversion keep quiet. You may see some people's work spaces as being rather chaotic and disorganised (P) as opposed to others who have a tidier, more organised workspace (J). Empathy and sensitivity may not be naturally forthcoming from supervisors with a preference for thinking (T) as they tend to be more objective and analytical – the majority of bioscientists show a preference for thinking (Blackford 2010). Lateral and 'blue-sky' thinking may come more easily to intuitive (N) types whose thoughts are more random than those of their sensing (S) colleagues, who are more practical and rely on factual information when formulating ideas. Working in harmony, N and S types can move their research forward by combining lateral ideas and explanations with real data and research findings.

Quote from Maria, PhD student

Now that I know I am an ESTJ, I don't despair when my supervisor complains that my analysis is too fragmentary. I actually would like to shout at him 'I am an ESTJ! What do you expect from me?!? You should be glad that I am gradually learning to focus on the big picture!'; but I just smile because being an ESTJ means that I usually get things done and this PhD shouldn't be an exception … ;-)

MBTI® and career development

Fundamentally, the MBTI® is a tool to enhance self-awareness. It is not about absolutes, it is about preference. To illustrate this, write your signature on a piece of paper. Then, change hands and repeat the process. Unless you are ambidextrous, you will get a good result and a bad result. The good result is from your preferred hand, the bad from your non-preferred. However, with practice you can improve your non-preferred performance. Thus, it follows that the preferred behaviours associated with your personality type can provide clues to your preferred ways of working and behaving. Equally, awareness of your non-preferred behaviours can help you to identify areas for personal development, e.g. if you have a preference for introversion you might want to work on your networking skills; if you have a preference for extraversion, you can work on your listening skills; if you have a preference for thinking, you can try to make a conscious effort to be more empathetic; Ps can learn to plan ahead more, etc.

Your personality preference can also offer clues about career areas which may be more suited to you (although bear in mind this is not a matching exercise and we can all do similar jobs but in different ways). Many studies have been conducted to determine which personality types are dominant in particular professions such as management, nursing, finance, social work and academic research. My own research (Blackford 2010), conducted on 150 bioscience postdoctoral researchers and doctoral students, showed that people enjoyed different parts of their research depending on their personality type. For example, those with an ISTJ preference were drawn more towards the technical aspects of their work, while ENTPs liked variety and going to conferences. ESTJ types were frustrated by the open-endedness of research and were looking to move into a career which was more goal and team orientated, while for those with a preference for INTP the opposite was true. The results also showed that the ratio of those with a preference for realistic facts (S) to intuitive types (those with a preference for conceptual thinking, N) was equal, whereas in the normative population this ratio is 3:1. Furthermore, there were five times more thinking (T) types than feeling (F) types in the bioscience population than in the general population (where the ratio is equal). This may account for the types of personality which are drawn towards scientific research, and it may influence the direction of your future career.

Awareness of your personality can also help you at interview, enabling you to demonstrate self-awareness while recognising areas for self-development (for example, when asked what are your strengths and weaknesses).

3. Values analysis

Although values play a major role in our lives, they do not tend to be articulated explicitly. Values generally don't come up in conversation and do not appear in job advertisements and yet they are fundamental to the way we run our lives. They influence our career choices in terms of the type of job, level of responsibility, work environment and organisation we are drawn towards. Schein (1990) cites the combination of personal values, competences and motives as the key to discovering your career anchor. Table 3.2 presents a summarised inventory, designed by Schein, which consists of statements describing eight career anchors. After filling it in and scoring your responses, you can refer to descriptions which equate to your two most important values.

Use the following scale to rate how true each of the items is for you:

- never true for me – 1
- occasionally true for me – 2 or 3 (depending on strength of feeling)
- often true for me – 4 or 5 (depending on strength of feeling)
- always true for me – 6.

Add up your scores for each career anchor and determine which one or two score the highest. Now read through the following descriptions of Schein's eight career anchors. Determine whether your score reflects which values hold the most importance for you.

Table 3.2 Values inventory (adapted from Schein 1990).

	Value preference	Career anchor
1.	I will feel successful in my career only if I can develop my technical or functional skills to a very level of competence.	TF
2.	I dream of being in charge of a complex organisation and making decisions that affect many people.	GM
3.	I am most fulfilled in my work when I am completely free to define my own tasks, schedules and procedures.	AU
4.	I would rather leave my organisation altogether than accept an assignment that would jeopardise my security in that organisation.	SE
5.	Building my own business is more important to me than achieving a high-level managerial position in someone else's organisation.	EC
6.	I am most fulfilled in my career when I have been able to use my talents in the service of others.	SV
7.	I will feel successful in my career only if I face and overcome very difficult challenges.	CH
8.	I dream of a career which will permit me to integrate my personal, family, and work needs.	LS
9.	Becoming a senior functional manager in my area of expertise is more attractive than becoming a general manager.	TF
10.	I will feel successful in my career only if I become a general manager in some organisation.	GM
11.	I will feel successful in my career only if I achieve complete autonomy and freedom.	AU
12.	I seek jobs in organisations that will give me a sense of security and stability.	SE
13.	I am most fulfilled in my career when I have been able to build something that is entirely the result of my own ideas and efforts.	EC

(continued)

Table 3.2 (*cont'd*)

Value preference	Career anchor
14. Using my skills to make the world a better place to live and work is more important to me than achieving a high-level managerial position.	SV
15. I have been most fulfilled in my career when I have solved seemingly unsolvable problems or won out over seemingly impossible odds.	CH
16. I feel successful in life only if I have been able to balance personal, family and career requirements.	LS
17. I would rather leave my organisation than accept a rotational assignment that would take me out of my area of expertise.	TF
18. Becoming a general manager is more attractive to me than becoming a senior functional manager in my current area of expertise.	GM
19. The chance to do a job my own way, free of rules and constraints, is more important to me than security.	AU
20. I am most fulfilled in my work when I feel that I have complete financial and employment security.	SE
21. I will feel successful in my career only if I have succeeded in creating or building something that is entirely my own product or idea.	EC
22. I dream of having a career that makes a real contribution to humanity and society.	SV
23. I seek out opportunities that strongly challenge my problem-solving and/or competitive skills.	CH
24. Balancing the demands of personal and professional life is more important to me than achieving a high-level managerial position.	LS
25. I am most fulfilled in my work when I have been able to use specialist skills and talents.	TF
26. I would rather leave my organisation than accept a job that would take me away from the general managerial track.	GM
27. I would rather leave my organisation than accept a job that would reduce my autonomy and freedom.	AU
28. I dream of having a career that will allow me to feel a sense of security and stability.	SE
29. I dream of starting up and running my own business.	EC
30. I would rather leave my organisation than accept an assignment that would undermine my ability to be of service to others.	SV
31. Working on problems that are almost unsolvable is more important to me than achieving a high-level managerial position.	CH
32. I have always sought work opportunities that would minimise interference with personal or family concerns.	LS

The values inventory and descriptions are abridged and modified from Schein EH. (1990) *Career Anchors. Discovering your real values*. John Wiley & Sons, Inc. Originally published by Jossey-Bass/Pfeiffer.

Technical/functional competence (TF)

Being an expert is central to what really pleases you about your job. It could be particular knowledge or technical expertise, which you have accumulated during your PhD and research, on which you would like to base your future career. As you move through your career you may be less satisfied with roles which require less specific expertise.

Technically/functionally anchored people want to pursue their area of expertise, preferring to become a functional manger as opposed to a general manager. Careers in science usually begin with some kind of technical specialisation. For TF types it is this role which they would prefer to perpetuate while others prefer to leave these functions behind. TF people like their work to be challenging where they are able to test their abilities and skills. Teaching and mentoring younger people can be a useful progression route for these people as being promoted into a general managerial position is viewed as undesirable as it forces them out of their specialty.

General managerial competence (GM)

Some people find that it is general management which interests them as they progress in their career. They have the range of competences which are required of a general manger and have ambitions to rise to the higher levels of the organisation where they will be responsible for major policy decisions and where their decisions will have influence. These people see specialisation as a trap and recognise the need to have knowledge of several functions. Key values and skills for these people are those that enable advancement to higher levels of responsibility with opportunities for leadership.

General managers tend to be able to make decisions under conditions where there is incomplete information. They can cut through the irrelevant detail to get to the heart of the matter and are capable of identifying and stating problems in such a way that decisions can be made. These people have the interpersonal skills to lead and manage people at all levels of the organisation. They have intergroup and emotional competencies too, but these competencies need not be developed to a high level as it is the combination of skills which is essential.

Autonomy/independence (AU)

AU people are those who feel restricted if bound by other people's rules, procedures and working hours. Regardless of what they work on, they prefer to do things in their own way, at their own pace and against their own standards. They find organisational life restrictive and intrusive into their personal lives. If forced to make a choice, they would rather give up the opportunity of a better job and remain in their present job if it offers more autonomy.

Most people require some kind of autonomy in their lives but for AU people this factor is overriding to the extent that they must be 'masters of their own ship' at all times. Research and development careers tend to promote autonomy and, if drawn to business or management, these people tend to go into consultancy or teaching.

Security/stability (SE)

Some people have an overriding need to organise their careers so that they feel safe and secure. Everyone needs some degree of security and, at particular life stages, it can be the major influencing factor, e.g. when raising a family. However, for SE people security and stability predominate throughout their careers to the point that these concerns guide their career decisions.

SE people are willing to give the career management responsibility to their employers and, in exchange for a permanent post, they are willing to be instructed in many parts of their career by their employer. For this reason they may be seen to be lacking in ambition but, in fact, many of these people can reach middle to high levels of management in (stable) companies. These people tend to favour the formal tenure systems which exist in the public sector but, of course, this system does not apply to postdoctoral research.

Entrepreneurial creativity (EC)

Some people have an urge to create their own business by developing new products or services or building new organisations. They are not necessarily inventors or creative artists; the creative urge in this group is specifically toward creating new organisations, products or services which can be identified closely with the entrepreneur's own efforts, and that will be economically successful. Making money is a measure of success.

Many people dream about building their own business; they may have expressed this dream as early as at school with small money-making enterprises. Such motivation may have come from their family which may have produced successful entrepreneurs. Many people want to run their own business due to an urge for autonomy/independence. However, people with entrepreneurial creativity need to prove that they can create businesses, which may mean sacrificing autonomy and stability. EC people tend to get bored easily and need to continually create and invent new products and services. They are restless and continually require new creative challenges.

Service/dedication to a cause (SV)

Some people enter certain careers because of central values that they want to embody in their work. These values are more important than the actual talents or areas of competence involved. Their career decisions are based on their desire to improve the world and make a 'difference'. Those in the helping professions, e.g. medicine, nursing, social work, teaching and religious work, are generally considered to hold this career anchor. However, dedication to a cause characterises other people who have entered their current research work in order to help people either directly or indirectly through their endeavours.

Service-orientated people need work that permits them to influence the organisations that employ them or move social policies in the direction of their values. They are driven by how they can help other people more than using their talents (which could be represented by many different areas of expertise).

Pure challenge (CH)

CH people are driven by challenge and the belief they can overcome any situation, any difficulty, solve the 'unsolvable'. They are determined, focused and competitive individuals. The military is an obvious career in which a CH person would be able to express their core values by taking on the enemy and centring their career on the service of duty to their country. However, other professions which are attractive to these people include research, management, sales and competitive sports (where winning is everything).

For these people, pay, promotion and the employing organisation are likely to be less important than whether the job provides opportunities to test themselves. Otherwise these people become bored, feel underused and even irritable.

Self-motivated CH people can be very single-minded, making life difficult for those working with them.

Lifestyle (LS)

Most people want to take account of their lifestyle when they consider their career choices but LS people will make this a priority, sacrificing opportunities in order to ensure their career is integrated with the rest of their life. Integration of lifestyle is an evolving function, so these people want flexibility in their work so that it moulds easily around their personal needs, e.g. moving only when it fits in with a family situation, being able to take on part-time work, day-care options, etc. This anchor was first observed in women but is increasingly seen amongst men, probably reflecting trends in society including the dual-career family.

If your career anchor is lifestyle, your work–life balance is the most important to you and you would not want to give up a situation which might upset the integration of your personal needs, family commitments and career. You may have to sacrifice aspects of your career (e.g. a promotion) as you place more value on your personal life than you do on work.

To illustrate the differences between career anchors, consider this scenario which I saw on a business opportunity programme. Two inventors had been motivated to set up their own company to supply creative, innovative teaching resources to school children in deprived inner-city areas. They were trying to procure investment in their company from a panel of venture capitalists. All those involved were entrepreneurial – the inventors and the venture capitalists. However, for the inventors, the main driver was the desire to help deprived children (SV). For the venture capitalists, the enterprise was only of interest if it could make money and become a successful business (EC), even though they were sympathetic to the cause. Those with an autonomous (AU) career anchor may also have a desire to set up their own company, driven by the desire for independence rather than the business *per se*. People's career anchors can change according to their stage in life. For example, for many people, lifestyle (LS) moves up the ranking when they have a family.

Conclusion

Having gathered all this information about yourself (skills, personality and values) as well as reflecting on other areas of your life (personal situation, interests, cultural and social background), how will this help in terms of your career management and planning? While there is no formula to match you to your ideal career, self-awareness is a crucial component of the career planning process. It can help you to recognise the kinds of jobs to which you are suited and help you to make informed decisions about your future career. Awareness of your areas of strength and weakness can offer clues about the kind of career development you need to undertake to improve your employability. Understanding yourself and your suitability for particular career sectors or jobs will also enable you to articulate your skills and motivations in an application or at interview (see Chapters 6 and 7). See career narratives 6, 8 and 12 (Appendix 1), in which self-awareness is cited as being central to discovering a preferred career path.

References

Blackford S. (2010) A qualitative study of the relationship of personality type with career management and career choice preference in a group of bioscience postgraduate students and postdoctoral researchers. *International Journal of Researcher Development* 1(4), 296–313.

Bolles RN. (2012) *What Color Is Your Parachute? 2012: a practical manual for job-hunters and career-changers: 40th anniversary edition.* Berkeley, CA: Ten Speed Press.

Council for Industry and Higher Education (CIHE). (2010) *Talent Fishing. What businesses want from postgraduates.* London: Council for Industry and Higher Education. Available from: http://bit.ly/tiFRHh.

Hawkins P. (1999) *The Art of Building Windmills. Career tactics for the 21st century.* Liverpool: GIEU.

Jung CG. (1923) *Psychological Types.* New York: Harcourt, Brace.

Law B, Watts AG. (1977) *Schools, Careers and Community.* London: Church Information Office.

NPA Core Competencies. Available from: www.nationalpostdoc.org/competencies.

Prospects Planner. *What Jobs Would Suit Me?* Available from: http://bit.ly/qmvRE1.

Schein EH. (1990) *Career Anchors. Discovering your real values.* San Francisco: Jossey-Bass/ Pfeiffer.

Schwartz SH. (1992) Universals in the content and structure of values: theoretical advances and empirical tests in 20 countries. In: Zanna MP (ed) *Advances in Experimental Social Psychology*, vol. 25. New York: Academic Press, pp. 1–65.

Super DE. (1981) A developmental theory. In: Montrose D, Shinkman C (eds) *Career Development in the 1980s.* Springfield, MA: Montrose.

Vitae. (2011a) *Coaching for Research in UK Higher Education Institutions: a review.* Cambridge: CRAC.

Vitae. (2011b) Researcher Development Framework. Available from: www.vitae.ac.uk/RDF.

Watts AG. (1977) Careers education in higher education: principles and practice. *British Journal of Guidance and Counselling* 5(2), 167–84.

CHAPTER 4

The job market

Where do you find job vacancies and how do you scrutinise them? How do you access information about jobs that are not being advertised? Which careers will suit you? Being aware of the types of jobs being offered and where to find them is a key part of the career planning process, as described in the DOTS model (see Chapter 2). Jobs can be found in the 'visible' and 'hidden' job markets (Box 4.1). This chapter provides information about how to access both.

As researchers and doctoral students working within an academic environment, it is likely you are well informed about academic research, funding and related opportunities. You may hear about these through personal contacts, via specialist discussion groups, during conferences or advertised on science job sites. However, as you look further afield to the wider job market, your knowledge becomes more limited. This is not surprising since you work in an environment surrounded by other researchers and, perhaps, rarely come into contact with professionals from other career sectors. A recent survey of doctoral graduates in the UK (Vitae 2012) revealed that over 40% of people had studied for a PhD with the intention of forging an academic career. In another survey (Eurodoc 2011), this figure increased to 67% when current doctoral students across Europe were asked about their career intentions.

As well as having very positive reasons to remain in academia, many researchers stay because it represents familiarity. Equally, a tenured position is seen by many as the 'prize' within university circles, with non-academic career options deemed by some to be second best or the result of a 'failed' academic career. However, unless you think you have the potential to realise a full academic career, it would be unwise to undertake too many short-term postdoctoral posts without considering other options, otherwise your career may end up at a 'dead end'. Researchers may be persuaded to stay for a further contract by their supervisor, who relies on them to generate data, or organise and manage their lab and group members. As stated by Dame Athene Donald FRS (professor at the University of Cambridge) in Rumsby (2011), 'There is a tension …. between the needs of the individual researcher and the needs of the PI who are under pressure to produce "outputs" … their line manager may have a vested interest in keeping them on in some useful role or other, rather than encouraging them to spread their wings and try out new things'. In addition, with statistical evidence revealing very limited success rates for securing a permanent academic post (see Chapter 1), aspiring researchers face very tough competition and are well advised to have a contingency plan in case things don't work out.

Career Planning for Research Bioscientists, First Edition. Sarah Blackford.
© 2013 Sarah Blackford. Published 2013 by Blackwell Publishing Ltd.

> ### Box 4.1   The visibility of job markets
>
> **The visible job market** is where the employer advertises their jobs using media such as the internet, magazines and journals. Here, it is up to you to approach the employer to make yourself known to them and 'sell' yourself using their specified method of application.
>
> **The hidden job market** is one where jobs are not advertised and may not even exist. It is up to you to approach the employer directly using a speculative application or by networking and making yourself more visible. More subtle approaches such as word of mouth and personal recommendations are strategies more closely associated with the hidden job market.

Career paths are not the linear trajectories they used to be and although actual figures are difficult to pin down, people are likely to have three or more careers in their lifetime. Job sectors rise and fall with the economic tide so that current occupational fields may disappear to be replaced by new ones over a period of just a few years. Even those in tenured academic positions may change jobs to work in industry or policy making or take up leadership roles in other organisations. Staying aware of the job market and maintaining your employability are advisable whatever your circumstances since we are all vulnerable to change and redundancy. For more information on this see Chapter 5 on personal and professional development.

Career sectors

As a research bioscientist, you are likely to be considering three main career options:

- an academic research career
- a career related to bioscience/science (see Box 4.2)
- a non-scientific career (see Box 4.2).

Academic career path (see Appendix 1: Career narratives 1–3)

Working 24/7 in the lab is not an effective way to develop a successful academic science career. Generating quality data and publishing peer-reviewed papers are necessary and important activities to demonstrate that your research is of the highest international standard within your field, but there are additional skills, responsibilities and experiences which are important if you want to rise up through the academic ranks. With the clock ticking away for a limited time during your temporary research contracts, you need to ensure that you are adding to your skill-set (and not just technical skills) to secure an academic position.

You have to show a level of independence in your research ideas so that you can set up your own research group and establish a niche. The ability or potential to formulate new ideas and demonstrate a degree of visionary thinking and then

---

Box 4.2   Career options for bioscientists

The following list includes examples of occupations in the bioscience and non-science sectors. Your skills and personal qualities will be marketable to both sectors (see Chapter 6).

Bioscience-related jobs

- Research and development (university/industry/research institutes/government)
- Clinical biochemist/immunologist/microbiologist
- Associated clinical careers: clinical trials manager, data manager
- Technical consultancy
- Product manager/technologist
- Specialist, e.g. toxicologist, pollution control, bioinformatician
- Patent attorney/agent/examiner
- Regulatory affairs/technology transfer manager
- Medical doctor/nurse/veterinary surgeon
- School teacher/college lecturer/tutor
- Science communication/journalism/publishing
- Science strategy and policy
- Research/data management and administration
- Scientific sales and marketing

Non-bioscience jobs

- Research (government/think tanks/policy/marketing)
- Accountancy/finance
- Self-employment/consultancy
- Project manager
- Sales and marketing
- Management, e.g. retail, operations
- Administration, e.g. university, government
- Librarian/information manager
- Legal services (lawyer, barrister)
- Human resources/recruitment services
- Conference organiser

---

secure funding is fundamental to leading a research team. Forming international collaborations, publishing and presenting papers, supervising and teaching will all contribute to your research profile. There are opportunities to enhance these skills during the course of your postdoctoral career so make sure you engage with them or you may lose out to those who do. Networking is invaluable for becoming well known within your research field and forming collaborations, which may lead to a jointly funded project or a job offer from another group or institution.

Bioscience- or science-related careers (see Appendix 1: Career narratives 4–17)

For careers outside academia, working 24/7 in the lab, again, is not an advisable strategy. Even if you are aiming for a research career, you need additional personal skills to support your technical expertise. Industry works as a team so evidence of capable communication, team working and a can-do attitude are paramount.

Beyond research, there is a plethora of other careers requiring different skills and personal aptitudes. For industry and other science-related careers, a science background or PhD will usually be mandatory. Employers want to see evidence of consistent output, coupled with the skills and experience required to do the job. A promising science journalist needs to show evidence of their writing ability. Aspiring patent examiners should demonstrate a high degree of accuracy, excellent written communication skills and commercial awareness. Entry into school teaching may require applicants to have had some experience working with school students through outreach work, demonstrations, workshops or school visits.

Careers outside science (see Appendix 1: Career narratives 18–20)

For some, leaving research along with their subject knowledge, scientific research-related skills and work environment is an attractive career option. Employers generally regard scientists highly, not because of their subject knowledge but for the skills they can demonstrate, e.g. numeracy, accuracy, analytical mind, problem solving, hypothesis testing, creativity and communication skills. Many researchers want to use these skills and are not concerned about remaining in research or even in a science-related work environment.

Choosing to do something completely different can offer a new challenge and new opportunities. However, don't underestimate the amount of effort involved in changing career paths. It is not straightforward so plan your exit strategy carefully. Think about how you can develop your career towards this new career path; undertaking extra activities, networking, work shadowing, even taking courses and further training which will enhance your relevant skills and increase your knowledge of this field (see Chapter 5). Find out who the major players in the job sector are and make tactful speculative enquiries.

Examples of job advertisements

How do you know which jobs are suited to you and how do you find out more about different career sectors? Researching the current job market is an effective way to find out about occupational fields and what is required for you to gain access to and be successful in them. By examining the job market you will (1) increase your awareness of the career landscape; (2) start to identify the kinds of careers which may be of interest to you; and (3) pinpoint gaps in your skills and experience which you may need to address if you want to be a realistic candidate. Bear in mind that if employers are looking for someone with a PhD, there is a reason for this. Recent surveys of large companies (Council for Industry and Higher Education 2010; Souter 2005; Vitae 2009) identified specialist subject knowledge, research expertise and analytical thinking as being significant qualities desired in PhD-qualified candidates.

The following section contains 15 vacancies with their job titles, corresponding descriptions and requirements. The job specifications are abridged and composite versions of real job adverts which were published in 2011. Examine the job descriptions and consider the following questions.

- What specialist knowledge and skills are required in each of the posts?
- Which post(s) interest you the most and why? Disregard the subject specialism of the academic research posts and focus on the job specification.
- How well can you match your current experience and skills against the requirements of these jobs (see Chapter 3 for a personal skills audit)?
- What action could you take to improve your suitability for these jobs?

(a) Postdoctoral posts

(i) Postdoctoral fellow (Denmark)

Applications are invited for a position as postdoctoral fellow for 2 years to work on proteomics of wheat tissues with emphasis on redox regulation and posttranslational modifications. The research concerns protein structure/function investigations, with a focus on carbohydrate interactions relevant for raw materials, food quality and nutrition.

Applicants should have a PhD in molecular biology or biotechnology or equivalent academic qualifications. The position will be offered to an applicant with a documented research record at an international level within proteomics and characterisation of protein posttranslational modifications. In the assessment of the candidates, consideration will be given to scientific production and research potential; the ability to communicate, promote and utilise research results; international experience; the ability to contribute to development of the department's internal and external co-operation.

(ii) Research associate (UK)

The postholder will be involved in the identification of new genes and genetic variants which cause early-onset Alzheimer's disease, based on next-generation sequencing. This post links into collaborative work in an International Alzheimer's Disease genomics consortium.

Applicants will need to have or be able to develop skills in biology, genetics and the bioinformatics of next-generation sequencing data. You should have obtained a PhD in a related field and will join an enthusiastic team of researchers in clinical, genetic and bioinformatics research. Ample opportunity for professional development, for example through regular meetings and conference attendance, will be provided.

(b) Tenured academic posts

(i) Lecturer (UK)

The Department of Biological Sciences has research strengths in environmental biology, molecular biophysics and molecular medicine. Applications will be particularly welcome from those proposing research in areas that enhance or complement the existing research activity of the Environmental Biology Research Group. Specifically, we are keen to attract a member of staff with interests in the application of the postgenomic technologies, systems biology approaches, proteomics or comparative genomics to address fundamental questions in plant or microbial biology.

Preference will be given to candidates with an established or developing record of research and a strong publication record commensurate with their experience. The successful candidate will be required to carry out high-quality research and obtain research funding, supervise postgraduate research students and carry out teaching and administrative duties.

(ii) Professor (Germany)

A W2 professorship has been established in the field of experimental molecular medicine. The position holder is expected to perform independent research in the field of human diseases. The Institute of Molecular Genetics research focus is in immunology. The ideal candidate will have experience and publications in the field of multiple sclerosis and will be able to integrate her/his research within the focus of the IMG, including the use of genetically modified mice already available in the institute.

The candidates will be chosen based on compatibility with existing research in the University and in the IMG. The candidate should provide evidence of past independent research that includes publications in high-ranking journals and external research grants. In addition, the position holder is expected to integrate with existing collaborative grant programmes already running or in preparation.

(iii) Assistant/associate professor (USA)

The Department of Biology invites applications for a senior faculty position in molecular, cellular biology of micro-organisms. The Department has research and educational programmes in cellular, genomic and molecular biology, as well as ecology. The university has recently built a genomics research facility that is fully equipped and dedicated to support interdisciplinary research in biotechnology and genomics of micro-organisms.

Applications are invited from individuals with well-funded outstanding research pro-grammes that complement our existing research strengths involving model organisms or micro-organisms of industrial, environmental or medical relevance. The successful candidate can be appointed at the rank of associate professor or professor in accordance with experience and is expected to participate in graduate and undergraduate teaching and collaborative research.

(c) Research in industry

(i) Research and development scientist (biotechnology company, USA)

Part of a highly motivated team of R&D scientists, you will design, initiate and conduct research and development projects including lab and field studies to support the insect, disease and weed control Product Evaluation R&D programme. You will provide technical marketing support for registered products in support of sales and marketing efforts.

Critical knowledge and experience
- *PhD in one of the agricultural sciences such as entomology, phytopathology or weed science with training and experience in applied sciences preferred.*
- *Experience conducting biological field trials to evaluate the efficacy and crop response of insect, disease or weed control products, and related experimental techniques.*

Critical technical, professional and personal capabilities
- *Excellent analytical and people skills. Skilled in interacting and communicating with internal and external customers.*
- *Excellent oral and written communication.*
- *Strong organisational skills to complete large field programmes.*
- *Successful interaction with peers and technical management.*

(ii) Molecular biologist (India)

The scientist will be responsible for developing a research plan for new projects, designing experiments and executing them with the assistance of associates and collaborators. The primary responsibility will be to design and execute research projects in the area of vector construction, tissue culture and molecular characterisation of transgenic plants. The successful candidate will train and manage research associates and assistants to conduct molecular experiments and work at the bench to demonstrate or establish new techniques. S/he will ensure safe working practices and work towards the improvement of protocols and working conditions. S/he will work closely with other functional groups within the company to effectively share knowledge.

Qualifications, competencies and experience
PhD in molecular biology or related field with 2–4 years of postdoctoral experience in a reputed lab working on molecular biology projects. Competencies include:

- *ability to conceptualise new projects and develop research plans to achieve the stated objectives*
- *demonstrable scientific competence as evidenced by peer-reviewed publications/patents*
- *people skills to lead associates and establish fruitful collaborations*
- *excellent interpersonal, oral and written communication skills*
- *good organisational and multi-tasking abilities*
- *ability to meet deadlines.*

(d) Science-related jobs in industry

(i) Regulatory affairs manager (Germany)
Our client provides support in regulatory affairs for company product portfolio within a local operation.

Responsibilities
- *Operational regulatory affairs activities.*
- *Maintenance of labelling information to internal and external partners.*
- *Communication to health authorities, regulatory authorities and institutions.*
- *Supporting internal global RA, other R&D departments and collaborative support for commercial teams.*
- *Ensuring adherence to applicable laws, codes and regulations both global and local.*

Requirements
- *Bachelor's degree or equivalent experience required.*
- *Preferred scientific/healthcare field.*
- *Significant experience of drug development within a regulated environment.*

(ii) New products outreach lead (large biotechnology company, USA)
We are looking for a highly motivated scientist to join the regulatory policy and scientific affairs team to lead the scientific outreach strategies for new products. The successful candidate will interact broadly with teams across the company's technology, regulatory, government, industry and media functions to engage and inform regulatory authorities, government institutions and other stakeholders to improve understanding of the science, technology and safety of our new and innovative products.

Responsibilities
- *Planning outreach initiatives and research studies.*
- *Facilitating publication of peer-reviewed journal articles.*
- *Organising workshops, symposia and other meeting venues.*
- *Developing communication resources appropriate for different stakeholder groups.*

Required skills/experience
- *PhD in molecular biology, biochemistry, genetics or related field.*
- *Strong networking and relationship skills and the ability to develop a network of international experts in key disciplines.*

- *Proven record of leading multifunctional teams, developing strategic plans and delivering results.*
- *Demonstrated effective communication skills.*
- *Direct experience in the biotechnology or pharmaceutical industry.*
- *Knowledge and experience of regulatory science and biotech safety assessment, and an understanding of international regulations.*

(e) Science publishing and writing

(i) Assistant editor (USA)

Members of the editorial team evaluate manuscripts, oversee the peer review process, commission and edit secondary materials. The successful applicant will attend scientific meetings and visit laboratories to maintain contact with the international scientific community. Excellent communication skills and a willingness and ability to learn new fields are a must. Applicants should have completed a PhD in the biological sciences.

To apply, an interested candidate should submit a curriculum vitae, a short (500–1000 words) news and views-style article on an exciting and newsworthy recent development in biotechnology, and a cover letter explaining their interest in the position.

(ii) Technical author: biotechnology (UK)

As the technical writer/author, you will be accountable for producing user guides and other end-user documentation relating to new cutting-edge bioinformatics products and software applications. Collaborating with R&D groups, scientific software teams and business development consultants, you will be expected to utilise your technical expertise in the production of detailed technical documents for some of the UK's research professionals. This opening would be ideal for a bioinformatics/technical researcher looking for a change in career or a more diverse, multidisciplinary role.

The successful candidate will be educated to MSc or BSc level (PhD preferred) in a bioinformatics, statistics or life sciences discipline. You will have a real passion for technical disciplines with prior experience in writing, editing or documenting technical or scientific subjects. Previous experience as a technical author is not essential but a good technical background in business intelligence, database administration, data analysis or manipulation would be highly advantageous.

(iii) Science communications manager (UK)

You will provide the Institute director and group leaders with scientific writing, expertise and administrative support. You will write, edit and design a wide range of publication materials to communicate its research activities to diverse audiences. These would primarily consist of manuscripts for journals, abstracts, journal/grant reports and the annual report.

You will be a motivated scientific professional with highly developed and proven writing skills and financial experience. Reporting to the Director, you will be responsible for composition of scientific documents and managing a portfolio of research awards ensuring systems and procedures are in place to ensure compliance with funders' terms and conditions.

You should have a detailed knowledge of biochemistry and molecular biology and preferably laboratory experience. Good communication skills and the ability to work as part of a team are vital as interaction with group leaders and research staff is a key aspect of this role. You must be highly organised with good interpersonal skills, numerate with excellent IT skills and have strong written and spoken communication skills.

An important aspect of the post is the responsibility for co-ordinating research funding applications and experience in pre- and/or postaward research funding administration would be an advantage. You should be able to work under pressure and to challenging deadlines.

(f) Administration

(i) Alumni relations officer (Germany)

The alumni relations officer will manage events, programmes and initiatives designed to serve the needs and interests of our alumni and staff. Responsibilities include overseeing on-campus events as well as many small-scale on- and off-campus activities and gatherings; recruiting and directing volunteer leaders for local chapter meetings and other projects. The officer is responsible for updating and maintaining the alumni website, contributing to the annual newsletter and supporting the Alumni Association board in organising their meetings, preparing documentation and following up with initiatives, ensuring activities adhere to the Association statutes.

Qualified candidates will have a university degree and a minimum of 3 years' related experience. Knowledge of alumni relations is an advantage but not a must. Candidates should have excellent communications and networking skills, demonstrate strategic thinking and problem-solving skills and have an understanding of working in a complex international environment (preferably an academic environment). A proven track record of working with and motivating volunteers is a must, and the candidates will need a good knowledge of a data management program. Fluency in English is essential, as is spoken and written knowledge of German. Fluency in other languages, especially French, is desirable.

(ii) European research administrator (UK)

This post is for a full-time research co-ordinator, tenable for a fixed term throughout the combined lifetime of both grants. You will have responsibility for the day-to-day operation of the projects (e.g. meeting organisation, financial administration), as well as contributing to project management (e.g. strategic planning, quality management).

You will have a first degree in a scientific or numerate discipline, and will preferably have prior experience of European Commission regulations and funding mechanisms. You should be able and willing to undertake occasional travel within Europe.

(g) Management programmes

(i) Government finance office (USA)

The Finance Career Programme recruits talented individuals at entry level and prepares them through training and rotational assignments to become federal financial professionals in financial management positions located throughout the United States.

Entry requirements
- *Master's or higher degree.*
- *Teamwork skills and leadership potential.*
- *Commitment to a career in public service.*
- *Solid analytical, problem-solving and decision-making skills.*
- *Good communications skills – written and oral.*
- *Research and evaluation skills.*
- *Academic accomplishment .*

(ii) Graduate trainee management scheme (UK)

A fantastic opportunity has arisen for a team of graduates to join a leading management consultancy. Ten positions are available and for the right ambitious driven graduates, this is a fantastic opportunity with excellent earning potential. This role comes with lots of responsibility and the graduate will be responsible for the full sales cycle; the graduate will be following up leads from marketing campaigns whilst also proactively researching and identifying our client's market place in order to identify and generate new business opportunities. This role also includes client management and you will be tasked with managing a range of business relationships, nurturing them and building rapport with key contacts.

For this excellent opportunity we require the following.

- *Educated to degree level or equivalent in any discipline.*
- *Driven and ambitious with the desire to succeed.*
- *Excellent verbal and written communication skills.*
- *Self-motivated and confident.*

Analysis of job advertisements

Referring to the checklist of questions in the previous section (page 35), this type of interrogation is necessary whenever you are scanning the job market. It will increase the effectiveness of your job search and give you a more in-depth insight into careers that are less well known to you. When scanning job sites and other vacancy sources, it is tempting to glance over job titles or the first few lines of a job description and dismiss them if they do not contain familiar words. Job titles are highly variable so they are difficult to decipher properly unless you have a closer look at them (see career narrative 17 in Appendix 1 for more advice on this).

You may be used to focusing on a specific scientific discipline which you are knowledgeable about and which is of particular interest to you. However, as you move further away from education and study, it is necessary to look beyond subject discipline and qualifications alone. The job vacancy examples listed above demonstrate the extra experience, skills and personal qualities employers want. This information gives you clues to which kinds of jobs and roles are potentially suited to you. Therefore it is advisable to have a similar internal checklist of questions to ask yourself so you can assess which jobs are potentially interesting. You can also use the checklist to identify areas in which to focus your personal and professional development. In order to do this effectively, you need well-developed self-awareness (see Chapter 3). Here is a brief analysis of the 15 example job vacancies summarising their fundamental descriptions and requirements.

Research (academia)

You will probably be most familiar with job descriptions a, i, ii and b, i–iii although be aware of the different terminology used for the job roles (see Chapter 1, Box 1.1, which lists different terminology for 'postdoctoral researcher'). Countries use different conventions for titling their academic posts. In the UK a lecturer is a higher level position than in the US, where the equivalent position is more aligned with 'associate professor'. More often than not, a lectureship in the US is

> **Box 4.3 Habilitation**
> (by Dr Teresa Valencak, Veterinary University of Vienna)
>
> 'Habilitation' refers to an advanced doctoral degree for scientists in academia about 4–6 years after their PhD and exists in countries in Europe which include Germany, Switzerland, Austria, France, Russia, Poland, Hungary and Slovakia. It is awarded to applicants after an evaluation process during which the candidate needs to show she/ he has proven to be independent in their research, can publish, teach and acquire third-party funds. In other words, it aims at qualifying a postdoctoral scientist for a professorship position. Once the evaluation process is positively completed and the candidate has given a public lecture at the university, the 'venia legendi' is granted. This document is often also required to officially and independently supervise and evaluate graduate students. Although mostly no longer obligatory, and despite being a relic from former times, 'habilitation' greatly helps to secure a position in science as it provides advanced scientists with a certificate of all professorial duties for an independent position in academia.

a temporary teaching post, while in the UK it usually carries tenure. 'Assistant professor' is a more junior post than 'lecturer' in the US and in some countries in Europe it is necessary to undergo a process known as 'habilitation' before being accepted for a permanent professorship (see Box 4.3).

The following website has extensive resources including roadmaps of academic career paths in Europe: www.ucl.ac.uk/careers/researchers/resources. Career narratives 1–3 (Appendix 1) also compare and contrast academic posts in the US and UK.

Advertisements for postdoctoral positions focus primarily on the research methodologies and technical know-how required to carry out the research project to generate data (a, i, ii). Many doctoral students apply for their first postdoctoral position, and subsequently a second, with relative ease since their skills are usually a close match for the role. However, as posts become more senior and develop into permanent academic positions it is clear (b, i – iii) that the institution is seeking more experienced researchers or established academics. Evidence of research independence, an international profile, capacity to raise funds and leadership potential are required, or implied. Note that one of the postdoctoral adverts (a, i) is seeking candidates who can demonstrate the 'potential for academic excellence', i.e. you would not be expected to have achieved this already. Rather, you need to provide proof of your potential to succeed as an academic. Evidence such as having successfully secured travel grant funding to attend a conference, being an active member of a learned society, collaborating and sharing your data and having a reputable list of publications will all count in your favour. If, however, you do not find yourself in this position after a number of postdoctoral posts or fellowships, you may have to rethink your career ambitions. Perhaps you have not been lucky with your research results. Maybe you need to reflect on your motivations and consider alternative strategies or careers.

Research and technical (industry)

Specialised research and technical skills are essential for the industrial research vacancies which stipulate a PhD (c, i, ii). However, there is more emphasis on interpersonal skills: the ability to interact and collaborate with other groups, motivate and lead a team, train others and be team orientated. You should be able to provide evidence of personal skills developed within or outside your professional role. Your research and associated experience are highly transferable and conducting basic research into the business (trends, clients, etc.) and providing evidence of commercial awareness gained from, for example, entrepreneurship competitions, short business courses or industrial collaboration will stand you in good stead.

For technical or service jobs in industry (d, i, ii), the work is usually quite specialised so previous experience of working in a similar setting is usually preferred. Furthermore, a scientific understanding of the work which is being conducted by the company means that a PhD or another specialised qualification may also be specified. Little or no interest in your publications, funding or academic research experience is likely *per se*, apart from it demonstrating that you have been productive in your research and can transfer your skills into an industrial setting. You will need to convince the employer about your motivation for moving into industry and that you are ready to project manage at a more general level and in a supportive or service role. Your CV will need to reflect this (see Chapter 6 and Appendix 3).

Science communication

For many science communication jobs, a scientific background is required but not always a PhD (although a PhD is evidence of a high-level scientific background). Many research bioscientists are interested in a career in science communication. But what is 'science communication'? It covers a very wide range of careers including publishing, science/medical writing, web editing, demonstrating science to the public/school students, media, press and public relations, journalism and education, amongst many others. The job vacancies (e, i–iii) demonstrate the diversity of this industry and you can also refer to career narratives 10–14 in Appendix 1 for more in-depth descriptions. Your interests, personality and values (see Chapter 3) will determine which types of science communication may suit you best. Fundamentally, however, if you are seriously considering a career in science communication you need to be able to demonstrate that you can turn complex scientific information into something more understandable to other audiences such as the public, schools, doctors, patients, science generalists, etc.

Building up a portfolio of evidence is important and you can do this during your PhD or postdoctoral research by engaging with activities such as writing for magazines, university newspapers, volunteering during open days, applying for internships and getting involved in science festivals. You could set up your own blog and provide digests of research in your field, for example, or offer to contribute to other websites (see Appendix 2 for more about using social media). For some job applications, you may be asked to bring your portfolio with you to interview or you may be set a writing task. Science communication lends itself well to remote working and so many science communicators take the opportunity to go freelance to ease their work–life balance (see Self-employment below).

Administration

The administrative posts (f, i, ii) do not require a PhD qualification but are seeking relevant transferable skills associated with these positions such as teamwork, communication, leadership and project management. Many science-related support roles exist in universities and other organisations which are attractive to those who want to move away from the bench but still be involved in science, and would prefer to remain within an academic setting. Some of these posts are highly specialised and require high-level knowledge and skills such as policy work, patent examination and senior administrative roles (see career narratives 9, 15, 16 and 17 in Appendix 1).

General management

Doctoral students and junior postdoctoral researchers are eligible for many graduate management training programmes or entry may be possible via an alternative fast-track career path. Attending career fairs, many of which are organised on university campuses, provides an opportunity to talk face to face with the representatives of large companies that run these programmes. If you take your CV, make sure it is adjusted towards this type of career. As you can see from the advertisements above (g, i, ii), these companies have little or no interest in the specific details of your research, degree discipline or publications. Most graduate management entry is via the company's application form so you will need to convert your research experience into transferable skills which demonstrate that you have the potential to succeed in management (see Chapter 6 and Appendix 3, CV4). Career narrative 19 in Appendix 1 describes a career in technology consultancy.

Self-employment

Although not present in the listed advertisements above, since self-employment is not generally advertised, I will cover it briefly here. Self-employment can mean a number of things: setting up a company or a partnership, doing consultancy work or going freelance. If you are considering any of these options the fundamental considerations will be:

- What expertise do you have to offer and to whom?
- Do you have a good network of contacts?
- Can you compete in the market and will people be willing to pay for your services/product?
- How will you get started?

Career narratives (8, 11, 14 and 18) in Appendix 1 chart the transition of a number of research bioscientists into freelance work and self-employment, including science writing and editing, setting up a company and a consultancy partnership. All of the people built up their expertise before they left their secure employment, taking their knowledge, skills and a network of contacts with them. Their motivations for setting up independently were different but, in most cases, were due to work–life balance issues.

General rules and advice about self-employment are available but information varies and is largely country specific so you will need to research this according to where you are based. Here are two websites to check for further information:

- www.prospects.ac.uk/self_employment.htm (UK)
- www.usa.gov/Business/Self-Employed.shtml (USA).

How do you find out about job opportunities and vacancies? As researchers, this should be a relatively easy task. Your own research project requires you to keep up to date with research findings and developments within your field of interest. You do this by reading papers, getting regular updates emailed to you, attending conferences and meetings and speaking to colleagues. You can use this well-developed set of research skills to conduct research into the job market. Knowledge of the career landscape will help to inform and guide you in your next career transition. You may be aiming for a postdoctoral position, lectureship or assistant professorship, in which case you need to be well informed about funding streams or openings in research groups operating in this field. If you are considering transferring your research skills into industry, the job market becomes less familiar and you are likely to need more time to research and find out about the types of jobs being offered, the kinds of skills and expertise required and where these jobs are being advertised. Other careers may be even more mysterious to you – you may not even know the range of options which are possible for you to consider. The final section of this chapter provides information on where to look for jobs in the 'visible' job market and suggests ways to extend your search using strategies and techniques to access the 'hidden' job market (see Box 4.1). Armed with a list of job sites and sources of information (Appendix 4) and a clearer notion of your own skills and strengths (Chapter 3), you should start to build a picture of the types of careers suited to you and how to gain entry to them.

Accessing the 'visible' job market

Websites

Jobs and career websites advertise general and sector-specific vacancies. Websites such as naturejobs.com, newscientistjobs.com and sciencecareers.sciencemag.org advertise scientific jobs and many also provide careers advice and information with case studies of people working in different professions. Euraxess and jobs.ac.uk focus on research positions, and specialist organisations advertise jobs in their sector using specific websites. Find out which websites are most useful to your job search and visit them on a regular basis or sign up to receive weekly job alerts. Be careful when nominating your keywords and the types of jobs you are seeking; you will miss out on potentially interesting jobs if you only provide narrow search criteria. A multitude of general jobsites populate the internet and it can be difficult to know which to target for legitimate jobs and reliable information. Use your university careers service to help you find genuine sites. Target the websites of professional bodies or contact them to obtain recommendations. Refer to the listed sources and further information in Appendix 4.

Organisations, especially large companies, universities and other public sector institutions, advertise current vacancies and provide information about their business on their websites. For specialised jobs, which are rarely advertised, you may need to find a contact to approach them on a speculative basis (see 'Speculative enquiries' and 'Social media' below).

Recruitment consultancies

Many companies use recruitment consultancies to hire new people or to head-hunt specialist individuals. If you are using recruitment consultancies, register with a number of them to increase your likelihood of finding the right job and keep in touch with them as they are at the interface between you and the employer.

Recruitment companies can be general or specialist. Their role is to carry out initial screening on behalf of a company to save them time and money. They will ask you to provide a CV or fill in an application form so it is important that you sell yourself and highlight your skills and abilities so that they can match you to a company's requirements. The recruitment consultancy may also conduct the initial interviews before candidates are put forward for further selection. Recruitment companies advertise their services on the internet so it is easy for you to find them and sign up. By networking with people in your target industries, you may discover their favoured agencies. Don't rely on this service alone, however, or you will restrict your job search.

University noticeboards and newsletters
Many research positions, internships, summer schools, courses, conference posters and other academic-associated opportunities are circulated around universities and posted on departmental noticeboards, so find out where they are and look at them from time to time. Most departments and universities issue a weekly e-newsletter to their staff and students which includes many of these opportunities so don't 'delete before reading' when they land in your inbox.

Magazines/journals
There was a time when most scientific jobs were advertised in a few key magazines and journals such as *Nature*, *New Scientist* and *Science*. Specialist magazines and job bulletins exist in other career sectors that you will need to research in order to find out where their vacancies are being advertised. Although these hardcopy media carry many job adverts, the majority of them are also published online. You may have to register first to access their vacancies.

Careers services
University careers services and employability centres advertise jobs and provide information about careers, much of which is posted on their website. Most are focused on undergraduate students but some universities provide services to researchers such as specialised workshops, advice, guidance and CV checks free of charge so they are worth a visit to find out what they have to offer.

Career fairs and conferences
Many companies wishing to recruit large numbers of graduates for major industries such as finance, management, administration, government and the armed forces exhibit at career fairs and conferences. However, these companies and small businesses rarely recruit for specialised research or technical personnel in this way since they do not require large numbers of such people.

Specialised science career fairs tend to be more biased towards engineering and technical companies. However, it is worth reviewing the exhibitors and even if only one or two are of relevance to you, it may merit a visit. Career fairs give you the opportunity to research companies and careers you are not familiar with or had not been considering. Many career fairs run a programme of talks to help you research job sectors further. Some even have careers advisers on hand to provide one-to-one guidance on aspects such as career choice, CVs and interview technique. A list of science career fairs can be found on my blog at www.biosciencecareers.

org/p/education-policy-careers-meetings.html, but this is not exhaustive so look out for local events or those being organised within larger conferences.

Professional bodies and organisations such as the European Science Open Forum (ESOF), One Nucleus, www.networkpharma.com and the National Association of Science Writers have networking opportunities, careers events and training sessions during their meetings. Look out for these and any travel bursaries to help you attend.

Scientific conferences

When attending conferences, it is worth signing up for supporting career and job programmes which have been organised for doctoral students and postdoctoral researchers. Research group leaders post jobs and look out for potential postdoctoral and senior researchers at these meetings. Careers talks and workshops offer you time to think about your career and cover subjects such as alternative careers, writing an effective CV, funding opportunities and science communication.

Accessing the 'hidden' job market

According to opinion on many specialised career websites and other sources (e.g. www.quintcareers.com/job-hunting_myths.html), only 15–20% of jobs are advertised. Even when they are, you may sometimes get the feeling that an 'insider' has already been lined up for the job. As well as extending your knowledge of the job market, you can increase your chances of securing a job by improving employers' knowledge of you and what you have to offer. Consider your current visibility (or celebrity!). How much do people know about you and your achievements?

When you apply for a job, your likelihood of success could be increased if you are already recognised as an expert in your field or a committed individual. Hiring is a risky business for employers so it is not surprising that they may be tempted to offer the job to someone they know and trust, or who has been recommended to them. Therefore, in order to tap into this hidden job market, you will need to make use of contacts and build relationships.

Speculative enquiries

Making speculative enquiries can be very productive. This is where a job vacancy does not yet exist and so you are looking for something that isn't there! Although not applicable to all career areas, many jobs are found this way. A well-worded speculative enquiry which lands on an employer's desk at the right time may prompt them to hire you rather than advertise the job. The most effective speculative approaches are made following thorough research of the organisation and job role. You can find people to contact in many ways. For example, if you see a job advertised which is at too high a level or not entirely relevant to your particular skills, but is within a company/organisation you think you would like to work for, keep a note of the details and any useful contacts. You can then use this information to approach the company on a speculative basis. For the best results, try to aim your enquiry at people who run the particular group or department you are targeting. Social media and discussion lists also offer opportunities to locate people to target speculative applications (see Appendix 2). See career narratives 6 and 7 in Appendix 1 to find out how two people used speculative applications to find a new job.

Word of mouth
Recommendations made and information obtained from people you know, or from contacts of people you know, are probably some of the most effective ways to secure a job. Many people cite contacts and chance encounters when asked how they got into their job or progressed in their career. Using contacts via your network will considerably enhance the likelihood of finding a job within the hidden job market.

Networking
Networking is one of the best and most productive methods for finding new opportunities. If you don't network you will be missing out on a huge part of the job market and your reactive approach may yield limited results. Nowadays there is much more access to people because of technology. You no longer need to be based in a particular place in order to network with influential individuals. That's not to say you won't be successful if you only seek and respond to job adverts – you may well have secured your current and previous positions in this way. However, as any good scientist knows, using a variety of methods to achieve your aim is preferable to using one method alone.

Many people are uncomfortable with the concept of networking as they see it as being false or a pretence at friendship. However, it simply means connecting with people, something we do all the time. If you look at your immediate network, it probably consists of a wide variety of individuals (Fig. 4.1). Fanning out from this, you can probably identify a more distant network comprising friends of friends and family, colleagues of colleagues, names you recognise within your institution but do not know in person, members of special interest groups and social networks. Beyond this, there may be people you would like to get to know but who you do not yet have any connection with. For example, there may be a particular research group you would like to collaborate with, or you may have identified a company you would like to work for but have no connections or personal links to them.

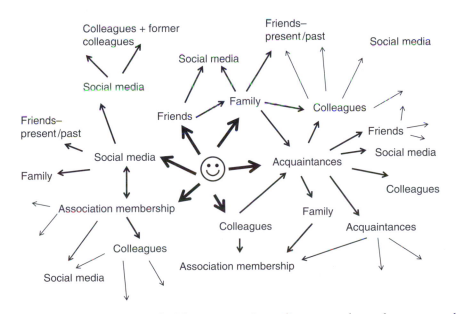

Fig. 4.1 Who is in your network? Those in your immediate network are also connected to other networks which will position you in an even larger matrix of contacts.

Perhaps you are considering a non-research career such as policy or publishing but you know of no one who works in this industry. You can make links directly or approach those who you think may be connected to potentially helpful people. If you have a rapport with someone, have demonstrated that you are capable, motivated and enthusiastic, they may go out of their way to try to help you with your career over and above someone they don't know. They may connect you with someone in their industry or invite you to visit their office or lab.

In addition to finding a way into a career sector, networking can also assist you in finding out about particular jobs and careers so you can test them out before committing to them. Those who are considering working in industry may wonder what the working environment is like and whether it would suit them. If you are thinking of changing your job completely into a new profession, this may be even more mysterious to you. Interviewing people already working in a career in which you are interested is one way to find out about it (although one person's view is perhaps not enough to rely on). Getting some real-life experience such as work shadowing, voluntary work or a short internship can be a more useful way to 'dip your toe in the water' to make a more accurate assessment as well as accruing further contacts. Many of the career narratives in Appendix 1 stress the importance of networking to secure a job within and outside academia (e.g. career narratives 2, 3, 10 and 18).

Making the most of your network

How do you go about networking? What strategies can you use to get to know people relevant to your current career or who might be able to help you with a career change? Networking requires a subtle approach. Sometimes it's more productive to go in through the 'side door' than through the 'front door'. For example, if you approach someone directly and unexpectedly, asking for a job, they can really only give you a 'yes' or 'no' answer. However, getting to know their career sector/industry/research area better, joining their wider network and connecting with people to whom they have links, as well as showing interest and knowledge of their industry, can produce more successful results. It shows that you have commitment and something to offer to the company and demonstrates personal qualities such as self-motivation and drive. The way you network and with whom you network depend on the types of careers you are considering.

It is likely that most of the people you know in your immediate network are working in academic research so they will be particularly useful to those wishing to forge a career in academia. Other doctoral students, postdoctoral researchers and more senior colleagues may be in networks of interest to you. Your supervisor/professor should be able to offer you guidance and opportunities to extend your experiences. For example, he or she will have links to other research groups, industry collaborators, policy makers, funders, etc. who could be useful to your next career move.

Extending your network is vital if you are looking to move out of academia. Your immediate network is less likely to have knowledge of people who are connected with this career area. Therefore, you will have to use a different set of tactics to get yourself known. This will depend on particular career areas so you may have to be quite creative.

Strategies for making the most of and extending your network are listed below. Note that there is overlap between this list and the one in Chapter 5 (Enhancing your employability), since networking is an important activity for personal and professional development.

Getting published

Publishing your work will get you known within your specialist field and may lead to you being invited to give a talk or poster within your department or at a conference; don't turn down any of these opportunities. A researcher from another group may be interested in discussing your data and offer to collaborate with you in the future – this will extend your network.

If you are interested in publishing as a career, you could target a publishing company (e.g. which publishes scientific journals with whom you may have had some contact) to enquire whether they offer internships or work-shadowing opportunities.

Organising events

Volunteering to organise seminars, events, a journal club, poster event, etc. within your own department will broaden your network and get you known. You may be given the opportunity to invite outside speakers to your institution (and even take them out afterwards) which will increase your profile and extend your network ofcontacts. If any of your colleagues or professors are organising a conference or meeting at your institution, offer to help out as this will also place you on the front line (see career narrative 2, Appendix 1).

Specialist groups

Joining a specialist group or learned society will bring you into contact with a whole community of people who are interested in your particular research field. Many learned societies have committees which you can join, including doctoral student early-career researcher groups. A bulletin (monthly, quarterly, etc.) will keep you up to date with the latest news and information about this field with features on members, forthcoming conferences, teaching and science communication resources and workshops.

Social media

Increasingly, social media are replacing other modes of networking online. If you are not already engaging with social media for work purposes, it is advisable that you start as soon as possible. Research-specific networks such as www.mendeley.com and www.research gate.net are probably well known to you. These sites allow you to connect with other researchers, share knowledge and increase your visibility. However, other social networks such as Facebook, Google groups, LinkedIn and Twitter are also being used by researchers to communicate and disseminate information. They can extend your network much further, so that you are swapping information with a wider sphere of people. These networks are also invaluable for those seeking careers outside academia and further away from their sphere of knowledge. Not only do they advertise jobs to their members/followers but they also provide information about people working within different job sectors. You can listen in to find out about hot topics and routes into a whole range of career areas. In particular, LinkedIn allows you to search for people and companies to make speculative enquiries. Research the background of employees who work for companies relevant to your job search to learn about their qualifications, skills and experience. This will help you to identify areas for personal or professional development and how to gain entry to this profession (see Appendix 2: Social media).

Discussion groups

Specialised discussion groups are a great way to network with your peers and they also advertise job vacancies relevant to their area of interest. When I advertise for a science communication intern each year, I post it on the psci-com discussion list (see Appendix 4) as I want to find someone who is already engaging in these types of activities. Specialist discipline and research groups are hosted by services such as www.jiscmail.ac.uk/ and are useful to those looking to stay within their research specialism. However, you will need to investigate other platforms to locate discussion groups of interest to you.

Conferences

Attending a conference in whatever capacity will always provide you with excellent networking opportunities. Popular names you have seen on journal papers will come to life and you can listen to them presenting their work and even meet them (or members of their research group). Conferences are a highly valuable forum for enhancing your PhD or postdoctoral experience, so make sure you get out and about at least once or twice a year.

Presenting a paper or poster at a conference will enhance your profile. A conference delegate may show interest in your work which could lead to other discussions which you can follow up later. These are all relevant reasons to engage with people to extend your network, but don't forget to make a note of people's names and contact details so you can reach them when you get home.

Many conferences hold networking events and specialist workshops. Try to attend these as they help you to meet people in small interactive groups. Many friendships and collaborations have been forged this way between people who would not ordinarily have networked with each other.

Many exhibitors and company representatives at conferences are former PhD students or researchers themselves who have changed careers. They may be able to give you advice or contacts to assist you in your job search, particularly in areas such as biotechnology, technical sales, marketing and publishing.

Your local network/university/institution

If you are considering a job in industry, your colleagues or professors may have contacts with collaborators with whom they can put you in touch. There may be commercial enterprises or spin-out companies based in your university whom you can approach.

For those considering science communication, offering to volunteer during school visits to your university or assisting those already involved in public engagement at your institution will extend your experience and bring you into contact with relevant people and their contacts.

Your university alumni office holds a database of graduates, some of whom will have agreed to offer advice and information to students and staff of the university, or even provide work-shadowing opportunities.

Members of staff who work at your university are specialists in a variety of areas such as finance, administration, international relations, public relations and so on and may be willing to help and advise you.

If you are interested in school teaching, local schools and teaching colleges are usually very accommodating and will allow you to observe their lessons and give you advice about the profession.

Approaching people

Taking a creative and proactive attitude is the fundamental key to success, so be prepared to approach people to make the first move. Below is a list of suggestions to help you to use and extend your network. However, take some time to think about other productive ways which might be more personal to you and the career in which you are interested.

- You may have to be quite creative to find useful new additions to your network. However, in any situation where you are looking to make use of your network, always ensure you have done your homework and are clear about what you want to talk about with your contact before you approach them (even if you know them well). It is advisable to email them first to establish your situation (but keep it brief) and to ask if there is a convenient time for you to ring or visit them so that you have their full attention.
- You can prepare for possible encounters during conferences by examining the programme beforehand and identifying professors and their group members who are likely to be attending. Alternatively, if you are interested in targeting a company which is exhibiting at the meeting, do some research beforehand so you make a good impression when you approach them to discuss careers in their industry.
- When you start networking, it is usually advisable to target less senior members of research groups or companies to begin with. If you establish a rapport with them, it will help you to approach higher level people.
- If you have arranged a meeting with a potentially useful contact, take the opportunity to write down your questions and motivation for your interest this career field. Having a short crib sheet to hand will ensure you do not forget anything vital (sometimes conversations can go off on tangents and you may need to refocus).
- As well as the conversation focusing on you interviewing your contact, be aware that you will be scrutinised too. Your contact will be assessing you professionally and personally, even if it is subconsciously, so it is really important that you make a good impression. You will not be expected to be highly knowledgeable about the subject under discussion if it is new to you, but use all the tips from the section on interview technique in Chapter 7 to ensure you give a good impression of yourself. If you come across as being hesitant or downbeat, it will create a negative impression. Therefore, wherever your conversation is taking place – at a conference, in an office, in the corridor – convey enthusiasm and be positive.
- Finally, keep your CV up to date and be ready to pass it on or email it to people with whom you have made contact. They may ask for it and a CV takes a long time to write if you plan to make a good job of it (see Chapter 6).

Conclusion

In this chapter we have examined just a handful of job vacancies and analysed the wide range of skills and aptitudes employers seek from applicants. We have also looked at where and how jobs are advertised (the 'visible' job market) and why and how they are not (the 'hidden job market'). Appendix 4 lists websites to help you find vacancies and information about career sectors in order to review the job market more widely and thoroughly. Getting an insight into other occupational fields should help you to prepare for your next career move. Enhancing your employability and developing your skills further will depend on the types of career(s) which interest you (Chapter 5). This will determine which activities you choose to engage with, the professional networks you join and the types of career development which will be of benefit to your next career move.

References

Council for Industry and Higher Education (CIHE). (2010) *Talent Fishing. What businesses want from postgraduates*. London: Council for Industry and Higher Education. Available from: http://bit.ly/tiFRHh.

Eurodoc. (2011) The first Eurodoc survey on doctoral candidates in twelve European countries. Available from: www.eurodoc.net/workgroups/surveys.

Rumsby M. (2011) *Research Careers – opening the discussion*. London: Royal Society. Available from: http://bit.ly/AyyAHZ.

Souter C. (2005) *EMPRESS: Employer' perceptions of recruiting research staff and graduates*. Swindon: RCUK. Available from: www3.imperial.ac.uk/pls/portallive/docs/1/45265697. PDF.

Vitae. (2009) *Recruiting Researchers: survey of employer practices*. Available from: www.vitae.ac.uk/CMS/files/upload/Recruiting_researchers_employer_survey_2009.pdf.

Vitae. (2012) *What Do Researchers Want to Do?* Cambridge: CRAC. Available from: www.vitae.ac.uk/policy-practice/167-515481/New-What-do-researchers-want-to-do-report-now-available.html.

CHAPTER 5

Enhancing your employability

The changing nature of work

Today, in our global economy, everyone needs to be poised to respond to changes to stay competitive. Organisations constantly evolve and adapt to their economic environment, creating an unreliable employment culture. As Meijers (1998) remarks, 'In advertisements the qualities required of future employees are more and more often described in terms of personal characteristics such as "dynamic", "prepared to change", "capable of carrying responsibility", etc.'. This means, he says, that people should be 'capable of dealing with work-related insecurity'. This uncertainty means that the workforce needs to be dynamic and flexible, possessing and developing the skills and knowledge required. Within the public sector and education, government policies change, funding is reallocated and different priorities give rise to redundancies in one area while bolstering another. Sometimes you can predict these changes and prepare for them; at other times they take you by surprise. It is within this ever-changing employment landscape that the most important skill we all need to possess is the ability to keep ourselves employable.

As highly qualified individuals with an extensive range of skills, you have the potential to apply for many job opportunities (see Chapter 4). However, your current expertise and skill-set will not sustain you throughout the course of your career. Developing yourself further and acquiring new skills and expertise to reach more senior levels and react to changing labour market trends is what will keep you employable. Defined in Chapter 1 as a set of personal aptitudes, skills and abilities which will enhance your chances of securing and maintaining employment, employability can be best maintained by taking a flexible and proactive attitude to your career planning. At times, your job may not be ideally suited to your personal preferences. You may even have periods of under- or unemployment. However, if you keep updating and developing your skills you can be optimistic that you will achieve a rewarding and enjoyable career.

Taking responsibility

Whichever career(s) you are considering, it is vital that you take ownership and responsibility for your career development and management. Researchers usually

Career Planning for Research Bioscientists, First Edition. Sarah Blackford.
© 2013 Sarah Blackford. Published 2013 by Blackwell Publishing Ltd.

know the end-date of their contract, which gives them an advantage over people who are made redundant unexpectedly with no time to prepare. However, some researchers do not plan for this inevitability. Instead, they wait until the end of their contract is imminent and, with little knowledge of their options, are ill prepared for their next career move and may be tempted to accept the first offer that comes their way. They may feel relieved when their professor/supervisor finds them more funding and extends their contract a bit further. However, this is a passive approach to career planning which is not sustainable and may lead to a worse situation in a few years' time when the research money finally dries up. You may be facing a period of flux yourself and finding it hard to cope with the idea of moving on in your career – this is a natural human reaction to change (see Chapter 8, Box 8.1).

You have the power to take control of your life and your career so that you are at the steering wheel of your own destiny, not passively sitting in the back seat. Do not rely on your supervisor to secure your funding again and again. This chapter lists some of the actions you can take in order to enhance your employability and increase your career prospects. You will probably be able to think of others more personal to you to add to this list.

Taking a proactive approach

Self-reliance and empowerment mean taking responsibility for your own development and career management, so that you are better prepared for your next transition. What you choose to do and the actions you take towards enhancing your employability will depend on your career goals. Even if you do not have firm ideas about your current career, you can still acquire experience which will enhance your CV and provide you with skills applicable to a future role.

In Chapter 3 (on self-awareness), you were invited to conduct an audit of your skills so you can present them in an application or articulate them at interview. The purpose of the exercise was also to help you to reflect on the wealth of experience and skills you have accumulated during the course of your career. There is plenty of scope to build on these skills and develop new ones to enhance your employability and increase your chances of securing your next position.

If you are aiming for junior postdoctoral positions, building on current research and technical skills will be the key to your short-term success. Securing a senior fellowship or academic post will require a more mature set of skills, such as demonstrable independence, ability to secure funding and an international research profile. For those wishing to transfer their research skills into industry, developing your team working, communication skills and commercial awareness will enhance your employability. If you intend to move further away from research to a new occupational field, demonstrating the transferability of your experience will be vital. In this case, your transferable skills will be more important to the employer than your knowledge and the details of your research project. Career narratives 2 and 9 in Appendix 1 relate how doing more and being proactive helped people's careers. Proactivity has been linked to high levels of self-awareness and adaptability – for more on this, see Chapter 8.

As well as taking a goal-orientated approach to your career planning based on theoretical models such as the DOTS model (see Chapter 2), chance, as explained in planned happenstance theory (Chapter 2), also plays a central role. Mitchell (2003) described planned happenstance as a way to convert unexpected events into career opportunities, requiring personal attributes such as being curious and optimistic, taking risks and being persistent.

Opportunities are more likely to be forthcoming if you are actively seeking them out rather than taking a passive approach. Many of the career narratives in Appendix 1 illustrate this point (e.g. career narrative 11), with people citing chance events contributing to their careers. These unplanned events appear to be the result of good luck but if you take a closer look, you usually find that the person was being proactive, making use of their networks or had been enhancing their skills so they could take advantage of an unexpected opportunity.

Personal and professional development

Regardless of whether you have a clear career goal in mind, have a few vague notions or no idea at all, personal and professional development will enhance your career prospects. Specific professional development may be necessary to enter a particular occupational field. Examining job descriptions and researching people working in your chosen career can help determine the experience needed to enter and succeed in this area. You may have to take a specialist qualification, do some initial voluntary work to build up a portfolio or demonstrate leadership ability. If you have no specific career in mind, you can still benefit from personal development to extend your current repertoire of skills and discover novel experiences, which may provide insights into other careers for you to consider. Furthermore, being proactive may lead you into undiscovered territories where chance events take you into a new career.

Here, I have compiled a list of some of the activities you could engage with to help increase your employability and employment prospects. You may already be doing some of these and more – the list is by no means exhaustive. You don't have to do all these extra activities but you will benefit from doing some. Consider whether you are making the most of opportunities available to you as a researcher working in a vibrant academic community with funds, facilities and a community of intelligent enthusiastic people around you. Some of these activities are particularly relevant for academic careers but all can offer valuable experience regardless of your career plans. Refer to the career narratives in Appendix 1 to determine the kinds of action which will enhance your employability according to different careers. Job advertisements and effective networking provide entry points into the 'visible' and 'hidden' job markets. Refer to Chapter 4 for additional information on the job market.

Practise making funding applications

Writing grant applications is directly relevant to academic career progression but also analogous to writing a 'business plan'. It demonstrates your ability to make a case for securing funding/investment and shows some degree of commercial awareness.

- Many postdoctoral researchers and doctoral students are prevented from applying independently for major funding streams, but this does not stop you investigating and becoming familiar with sources of research funding. Discuss with your supervisor how you can contribute to funding proposals to gain experience.
- Apply for small grants and bursaries. Universities, learned societies, funders and other organisations offer awards for travel, equipment and training to doctoral and early-career scientists.

Attend conferences and meetings

Directly applicable to enhancing your research career, contributing to and attending conferences also provide evidence of networking and communication skills.

- Get out of the lab. Go out and meet people, present your research and network with your peers.
- Apply for summer schools and other training workshops – they are usually free.
- Go to broader meetings or attend career development workshops.
- Offer to give a research talk in neighbouring departments in your university or if visiting a lab in another institution.

Collaborate

Research independence is vital if you are aiming for an academic career and collaborating with other groups is a useful way to steer your research focus away from your supervisor. Collaboration also demonstrates the ability to initiate networks and new projects.

- Research is no longer conducted in isolation. You are probably working in a research group with multinational, possibly interdisciplinary collaborators and with a wide range of stakeholders which could include business, non-governmental organisations, charities and the public. If you count up all those with whom you are (or have been) collaborating, you will probably find that you have a rich network of people who are involved in your research and who may be valuable contacts for your next career move. For example, if you are aiming for a research career in industry, ask an industrial contact if you can work shadow or spend some time conducting your research in their company. They may have contacts in intellectual property, regulatory affairs or patents if you are keen to investigate these avenues as a future career.
- You can extend your network by initiating your own set of contacts and collaborators. Arranging to visit other research groups to learn new techniques or use their equipment can act as a catalyst to find out about new research opportunities.

Join/set up a group

Taking the initiative to set up any new project will show you to be motivated and enterprising. Establishing a new group will also demonstrate your ability to engage people's interest and promote your ideas.

- If your department or university does not yet have a doctoral student or postdoctoral association, why not set one up? Approach your head of department for support – your department may have funds available to invite speakers or run social activities to enable you to network with each other. Guidelines for setting up a group are available (Vitae 2010). If a group is already in place, try to be an active member and play a role within it.
- Seminar series, journal clubs and other departmental groups also provide opportunities for you to get involved in research-associated activities. They generally provide excellent networking opportunities and can be a great way to meet high-profile visiting speakers. Academics and others who are managing them usually welcome assistance.
- Academic committees may require a doctoral student or postdoctoral researcher representative. This experience will enhance your knowledge of policy and administrative issues and bring you into contact with decision makers and other researchers in your department.

Join professional organisations

Being a member of a professional body or organisation associated with your area of interest will demonstrate commitment and broaden your perspective of the field by providing, for example, an overview of the latest news and policy developments or profiles of senior figures.

- Most professions have an associated organisation or body which professionals within it can join. There is usually a small annual subscription levied and, in return, members enjoy benefits such as a news magazine, reduced registration to meetings or training, access to other professionals in the organisation, job information and even professional recognition.
- Learned societies (see Box 5.1) offer a variety of benefits and you can join as many as you like! As well as preferential joining fees for PhD students and, sometimes, postdoctoral researchers, learned societies offer opportunities for you to write for their news bulletins, join committees, organise sessions or meetings and enter for competitions and awards. Many provide bursaries to fund travel as well as discounted conference registration fees, giving you the chance to present your work or visit other research groups.

Box 5.1 Learned societies

Learned societies are academic organisations committed to promoting their particular specialist academic field and supporting those who work within it. Their members consist of academics, researchers, postdoctorals and students and they sometimes extend their membership to include teachers and interested amateurs. There are dozens of bioscience learned societies worldwide. The larger, more active societies tend to be well funded from university subscriptions to their learned journals. For more information and a list of societies, see www.biosciencecareers.org/p/learned-societies.html.

Social media offer many opportunities to communicate to a much wider audience, extend your network and engage with professionals outside academia. For detailed coverage see Appendix 2.

Mentoring

Being a mentor is usually a voluntary duty and demonstrates that you are willing to take responsibility for someone less experienced. You can benefit from having a mentor yourself, who can give you an insight into areas of your profession that may not be accessible at your level. You will also receive advice related to your personal and professional development.

- A good mentor can be invaluable to your career. S/he need not be your supervisor; in fact, you may prefer someone unconnected with your research group so you can be more open and honest about your thoughts and plans. The mentoring relationship need not be part of a formal scheme. You can make informal enquiries to anyone you think could offer you the benefit of their experience.
- As well as having a mentor, being a mentor yourself is a valuable experience. As a postdoctoral researcher, you could mentor a junior postdoc or a doctoral student. A doctoral student may be able to mentor undergraduate or Master's students. Mostly, mentoring involves good listening skills and being able to offer advice and support, in confidence, to someone less experienced.
- National and international e-mentoring schemes exist, some of which are discipline specific or belong to professional bodies or learned societies. Others are aimed at specific groups such as women in science (see Appendix 4). There are many books and websites which can help you learn more about how to be an effective mentor. Universities and mentoring schemes should offer guidance and training.

Teaching

Teaching demonstrates your ability to convey complex scientific knowledge in a structured and comprehensive way. Your teaching techniques, assessment methods and pedagogical knowledge are marketable assets in themselves but also show evidence of communication, management and pastoral skills.

- If you are aiming for a career as an academic or college lecturer, it is worth getting experience of teaching Master's or undergraduate students during your early research career. Some university teaching policies may not allow research staff or students to teach but tutorials could offer a limited opportunity to get some experience. Once you get started, you may be able to enrol on an academic teaching course to gain a formal qualification, depending on national policy.
- For those considering school teaching, you can usually approach local schools to observe classes. Getting involved in school open days and outreach activities will also give you practice teaching school students.

Supervision

Being responsible for others' work gives you valuable experience of managing and advising team or group members. Supervision can enhance your interpersonal skills, such as giving instruction, providing constructive feedback and managing time and resources. In some instances it may be useful as evidence of leadership.

- Most postdoctorals have the opportunity to supervise doctoral, Master's degree or undergraduate students. Make sure you take up this opportunity as it demonstrates that you are able to take responsibility and manage others' work.
- Be careful to balance your supervisory responsibilities with your own research. You will still need to focus on your own project, so avoid being overloaded and consider the benefits to your career development.

Entrepreneurship

Being able to demonstrate innovative thinking, entrepreneurial skills and commercial awareness can enhance your employability, especially if you are aiming for a career in industry.

- Are entrepreneurs born or made? Review the following characteristics which make an entrepreneur and see if you can identify with any of them: creative, confident, communicator, autonomous, optimistic, team player, planner, risk taker, organised and passionate. You will be stronger on some of these attributes than others but all can be developed through experience and training. Your personal motivation for becoming more entrepreneurial may be your only barrier, although you may have limited opportunity to exercise some of these skills and personal qualities.
- Support and mentoring schemes are sometimes available in universities and institutes. Many universities possess knowledge transfer or intellectual property offices with staff you can approach to find out more or work shadow. Management schools may offer staff development courses and look out for specialised entrepreneurial associations and events being organised in your institution or nationally. See Box 5.2 for a personal perspective on entrepreneurship from a postdoctoral researcher.

Science communication

Without communication, no-one would know about your research. Communication skills are the most commonly specified amongst all job advertisements so enhancing these skills will always be an asset to your career development.

- Publishing is central to all academic researchers – it is your main 'output'. Try to gain academic writing experience during your PhD and, more critically, during your postdoctoral post as this is essential training for anyone aiming for an academic research career. Writing courses and workshops will help you to hone your skills and books on the subject can offer excellent support (see Appendix 4).
- Offer to write a review for a journal or broader academic titles. Not only will this give you a wider range of academic writing skills, it will also increase your knowledge of your research field.
- Communicating science to more general audiences is a skill, which you can improve during your PhD or postdoc. Science communication covers a wide range of styles (e.g. standing in front of a group of school students, blogging, taking part in debates and writing short articles). There are plenty of ways to get involved

Box 5.2 Entrepreneurship
(by Dr Xiaoqi Feng, postdoctoral researcher, UC Berkeley)

Researchers have many entrepreneurial qualities probably without even noticing them. Researchers are adventurous, and willing to take on risks and work under uncertainty; we are also original and innovative, which is essential for building something new; we are good problem solvers; and we have the determination and persistence to see things through.

What is entrepreneurship? For me, the definition of an entrepreneur is someone who commercialises an innovation, during which process he/she typically takes on financial risks while seeking profit. The three most important aspects of entrepreneurship are: the profit-generating commercialisation; innovation; and the involved financial risk. The innovation can be an idea, invention or a new technology. Entrepreneurship gives you an extremely valuable toolkit, not only for starting a business but which can be applied in many careers, including academia. In research, it is easy to get stuck in details and fail to see the bigger picture. Entrepreneurship teaches us to see if an innovation can make a difference, and to whom it is important, which is critical for a researcher to keep asking him/herself. Entrepreneurship also trains us in essential skills such as teamwork and communication. The time and resource constraints of entrepreneurship necessitate prioritisation and time management. What I find particularly useful in my case is the marketing skills I have learned. As an entrepreneur, one needs to market oneself, the team and the product, in order to attract financial investments and other support. To support a research project, a scientist needs to do exactly the same.

One effective way to apply science to the benefit of society is to commercialise it into products. That was what first got me interested in entrepreneurship. While I was doing my PhD, I teamed up with four friends at Oxford University and joined a UK competition – Young Entrepreneurs Scheme (YES Biotechnology). I learned a lot from this experience, met many interesting entrepreneurs, and truly enjoyed the team building and teamwork. Skills I learned from this experience hugely benefited my research, such as networking, presentation, teamwork and time management. Importantly, this experience further encouraged me to think in an applied way during my research.

The realisation of how much entrepreneurial thinking and skills training can help researchers urged me, during my postdoc at UC Berkeley, to found the Berkeley Postdoc Entrepreneur Program (BPEP) with three other bioscience postdocs. Since then we have offered a series of workshops training Berkeley postdocs and postgraduates essential entrepreneurial skills. We collaborate with business leaders for mentorship, provide a platform for team building and networking with funders and lawyers, and assist the start-up of companies. Around 150 now turn out at our workshops and we have seen teams being formed and funds attracted at our networking events. All of this suggests that entrepreneurship is a topic that is underaddressed in our scientific community, and that highly trained graduate and postdoctoral researchers, if provided with the right business training, can turn promising research into practical applications.

So how do you start thinking in entrepreneurial ways? As a researcher, to some extent, you already are. You probably do not even know it but you are creating new things, problem solving, communicating and collaborating with others, and marketing ideas and yourself to fund your research. All of these can be improved with some entrepreneurial training. Where to find them? Seek out entrepreneurial courses already available in your institute, join entrepreneur societies and competitions, and if there is none available, why not start one with other scientists, engineers and MBAs who share the same value and enthusiasm?

Box 5.3 Café scientifique

Café scientifique is a worldwide initiative which aims to bring science to the public in an accessible and enjoyable way. Usually held monthly in a designated social venue, such as a café or bar, the organisers invite a scientist to come and speak to the public on a topic of interest. People arrive early in the evening, eat a light meal and the speaker talks for about 20 minutes. Discussion and questions are usually chaired by the organiser. Cafés scientifiques are relatively easy to organise and are an opportunity for the organiser to meet a range of experienced scientists whilst developing his or her personal skills. There may already be a café scientifique in your town or you could consider setting one up yourself. For more information visit www. cafescientifique.org.

---

within and outside your university (see career narratives 12 and 13 in Appendix 1). Offer your university newspaper a science column, enter writing competitions, volunteer to write for learned society magazines, submit copy to web-based publications, start a blog (see Appendix 2: Social media) or set up a public event such as a café scientifique (see Box 5.3).

- Events such as science festivals provide a platform for scientists to showcase their research to the public in an accessible way and many advertise for volunteers to get involved during their events.
- For those thinking of pursuing science communication as a career, you can join specialised groups and discussion lists to network with professionals, improve your skills and find out about courses and competitions. For more information see the further reading section in Appendix 4.

Attend courses

Summer schools and technical workshops help keep you up to date with new techniques and are opportunities to meet like-minded researchers. A specialist qualification may be required for you to enter a new profession. Alternatively, you may need more general personal development. These courses can enhance your employability but you need to continue practising them in 'real life' so you don't forget what you have learned.

- Graduate schools normally provide career development programmes for their doctoral students. University staff development courses are generally open to postdoctoral researchers. Take advantage of these courses – they are usually free!
- Depending on the country in which you are based, university careers services offer specialised advice for bioscientists at all career stages. Workshops and one-to-one guidance for PhD students and postdocs may be available for those aiming for careers within and outside academia. Visit your careers service and see what they have to offer you.
- Other courses and summer schools provide training, ranging from specialist scientific techniques and methodologies through to more generic courses on management and leadership. Bursaries and grants are sometimes available to cover accommodation and travel expenses.

> **Box 5.4   Volunteering, placements and internships**
>
> Don't underestimate the power of volunteering or taking a short-term placement or internship. Some are paid and some are unpaid (or only pay your travel expenses). Some people see this as exploitation but if you are discerning, you should avoid this situation. Look at each opportunity carefully and ensure that it is something which could enhance your career prospects. Depending on your career goals, volunteering or undertaking an internship can be more valuable than acquiring a qualification. Courses are expensive and can be quite lengthy. They don't always give you the kind of training which a voluntary post can do, neither do they necessarily put you into contact with people in the profession.
>
> Choose volunteering opportunities which are aligned with your career interests and which will move you forward. Internships and placements are advertised in many different ways – formally on websites, discussion lists and social media or informally by word of mouth and contacts. You will need to be quite creative, use your network and perhaps apply speculatively to find out about these opportunities.

- Enrolling on a specialised course can boost your employability, depending on your career plans. Evening courses in subjects such as languages, computing and finance are usually offered by adult education colleges. One-week intensive or part-time options may be viable if you can balance your workload to accommodate them. In some cases, a Master's degree may be helpful to change career paths (e.g. information technology, bioinformatics or project management). Whether further courses will be advantageous will depend on the career you are pursuing and your existing skill-set.
- You can benefit from further training whatever career you are pursuing, as illustrated in career narratives 2 and 15 in Appendix 1.
- A list of training providers and career development events can be accessed on my blog: www.biosciencecareers.org/p/education-policy-careers-meetings.html.

Voluntary work

Voluntary work can be a springboard for changing direction into a new career, depending on the sector (Box 5.4). It helps you test out a new career and develop relevant experience and skills. It also enables you to meet and network with people you may want to work with or who may be willing to be a referee or write a statement in support of any subsequent applications you make in this field.

- Voluntary work can be done in conjunction with your current job or you may need to negotiate to take time off to do an internship or placement. For example, for academic progression it may be useful to volunteer to take on some teaching duties. You will increase your chances of getting into careers such as science communication if you have been doing outreach or science writing.
- Formal internships and placements (paid and unpaid) are more finite and formalised. Competition for them can be tough and so, even here, previous voluntary work or evidence of your commitment will help your application.

- Career narratives 12 and 15 in Appendix 1 illustrate where voluntary work has played a part in enhancing people's employment prospects.

Get support

Don't try to tackle your career on your own. As well as activities such as communicating, networking and mentoring, you can get support from local, national and international organisations.

- If your university has a doctoral student support group or postdoctoral researcher association, it is well worth joining as it acts as a mutual support mechanism and provides careers support and social activities. These associations can also act as a talking shop for research where you may find new collaborators or discover new information which might contribute to your own research programme (see 'Join/set up a group' above).

Examples of support organisations

BioDocs Lyon brings together young scientists in biology and aims to improve their employability in the public or private sectors. We organise scientific and socio-professional activities such as panel discussions about careers in life sciences and visits from biotechnology companies are organised each month. Because developing a network is the key to finding a job, BioDocs also takes part in the organisation of the annual BIOTechno forum which gives young researchers the opportunity to meet professionals in the private sector. *Dr Nisrine Falah, BioDocsLyon President (2011–2012), University of Lyon, France*

Our graduate association in the Biology department is run by three graduate representatives who also represent PhD students' views on departmental committees and support them if they have any other questions or problems. As well as these more formal activities we focus on the social and community aspect of student life, with an emphasis on preparing them for what lies ahead in their careers. In addition, we run a monthly event called GradShare (Friday, late afternoons). Free pizza, snacks and drinks are provided and we have an invited guest who talks for half an hour about their career path. Speakers have included professors, intellectual property lawyers and leaders of not-for-profit NGOs. We also organised a day of team-building games at the beach last summer which was part-funded by the department. *Toby Hodges, PhD student, University of York, UK*

- National and international support organisations are listed in Appendix 4. Many are well established (e.g. Vitae, the National Postdoctoral Association). They hold many invaluable resources on their websites which are free of charge. They also organise workshops and training events to support doctoral students and postdoctoral researchers.
- Specialised associations aimed at women in science offer fellowships for women returners, mentoring and a mutual support network. In many cases men are also able to take advantage of these opportunities. See Appendix 4 and Box 5.5.

Box 5.5   Women in science

by Dr Tennie Videler, co-chair of the Cambridge Association
for Women in Science and Engineering: www.camawise.org.uk

Women are underrepresented in employment in science, engineering and technology (SET), both inside and outside academia. Even in the biosciences, where women make up half the undergraduates, they still account for only 15% of professors. In all SET subjects, qualified women are not retained in similar proportions to men with the result that women are severely underrepresented in senior positions. For example, among SET academic faculty in the US in 2003, women comprised 18–45% of assistant professors (26% lecturers and 18% senior researchers/lecturers in the UK in 2007–08) and 6–29% of associate and full professors (9% in the UK in 2007–08). Not just in academia but in general SET occupations, fewer women with undergraduate SET qualifications enter SET professional or associate professional occupations (Committee on Gender Differences in the Careers of Science, Engineering and Mathematical Faculty 2010). Possible reasons include:

- *unconscious bias*: although few women experience overt discrimination, there is evidence that women (and members of ethnic minorities) are disadvantaged by unconscious bias (Steinpreis et al. 1999)
- *family responsibilities*: women still shoulder the majority of family responsibilities. Policies or informal practices can have a disproportionately deleterious impact on women (Kirkup et al. 2010).

Learned societies (e.g. Biochemical Society, American Society of Plant Biologists and Society for Experimental Biology) run initiatives and organise networking events at their conferences specifically aimed at women in science. National and international organisations and initiatives exist to support women in science (see Appendix 4).

Record your progress

Whatever personal and professional development you decide to undertake, keeping a record of progress will help towards your career planning.

- Reflecting on your skills and personal qualities (see Chapter 3) allows you to monitor your development during the course of your career so that you keep it moving forward. This will help you avoid complacency and stagnation and ensure that you are improving your employment prospects.
- A structured personal audit enables you to record your activities, and local and national personal development planning tools are available to assist you in the process (see Chapter 3).

Conclusion

The "job for life" no longer exists, which removes the security that people used to enjoy in the past. However, waiting for people to leave or retire in order to progress up the career ladder is not very inspiring. The flexibility and adaptability of careers these days can represent a much more exciting prospect. Taking a positive attitude and maintaining your employability so that you can respond to changes in your circumstances will put you in a strong position. You may need to revise your career ambitions if you wish to stay in a particular geographical area and, equally, you may have to compromise your personal life in order to take up a new post far away from where you are currently living. Wherever you find yourself in your career, never relax and think that you have nothing more to do. Continuing professional development is the key to enhancing your employability at every stage of your life – even in retirement!

References

Committee on Gender Differences in the Careers of Science, Engineering, and Mathematics Faculty; Committee on Women in Science, Engineering, and Medicine; Committee on National Statistics; National Research Council. (2010) *Gender Differences at Critical Transitions in the Careers of Science, Engineering and Mathematics Faculty*. Washington, DC: National Academies Press. Available from: www.nap.edu.

Kirkup G, Zalevski A, Maruyama T, Batool I. (2010) *Women and Men in Science, Engineering and Technology: the UK statistics guide 2010*. Bradford: UKRC.

Meijers F. (1998) The development of a career identity. *International Journal for the Advancement of Counselling* **20**, 191–207.

Mitchell K. (2003) *The Unplanned Career. How to turn curiosity into opportunity*. San Francisco: Chronicle Books.

Steinpreis RE, Anders KA, Ritzke D. (1999) The impact of gender on the review of curricula vitae of job applicants and tenure candidates: national empirical study. *Sex Roles* **41(7–8)**, 509–28.

Vitae. (2010) *A Guide to Research Staff Associations*. Cambridge: CRAC. Available from: www.vitae.ac.uk/CMS/files/upload/UKRSA_Guide2010_Dec13.pdf.

CHAPTER 6

Making applications

How is your personal branding? What have you got to offer? Are you promoting yourself at your best? No matter how experienced you are, or how well suited you think you are to an advertised job vacancy, you will still need to 'sell' yourself to others to convince them of your talents. Making applications is a major activity associated with the 'Transition' stage of the DOTS model (see Chapter 2). It is a crucial part of the career planning process and serves as your personal 'advertisement'. During face-to-face networking (see page 51) you promote yourself by speaking convincingly and using positive body language. Doing the same in writing is not an easy task. Applications serve one central purpose: they are a marketing tool to enable you to progress your career. This could be to secure a job within or outside academia, it may be an application for a fellowship, a summer school or Master's degree course. Whatever its purpose, you need to get it right first time!

Employer perspective

The employer has identified a gap in their workforce and needs to fill it with the most appropriate person. How does s/he find that person? Employing people is a risky business – if the employer gets it wrong, it can prove costly. Equally, funding bodies and course co-ordinators want to select the most deserving candidates for their grants or course places. Therefore, they look for applications that demonstrate most clearly that the applicant can fulfil the majority of the responsibilities and duties specified in their requirements (see Box 6.1). Furthermore, they want to see evidence that the applicant has some knowledge of their company or operations and has taken the time to find out something about them.

The importance of researching and analysing the requirements specified in the job description, and then matching your application as precisely as possible, cannot be overestimated. If you are aiming for internal promotion, analyse the higher grade responsibilities and expectations. As with all well-executed projects, preliminary research is always time well spent. Successful applications will be those which match the job description or funding specifications, showing a high degree of understanding of the criteria and knowledge required. This takes time and effort so it is better to focus on a few well-targeted quality applications than to send out many generic ill-matched applications. For electronic submissions, which ease the

Career Planning for Research Bioscientists, First Edition. Sarah Blackford.
© 2013 Sarah Blackford. Published 2013 by Blackwell Publishing Ltd.

Box 6.1 Job specifications

Employers and funders specify their requirements by listing 'essential' and 'desirable' criteria in their advertisements and announcements. However, the majority of applicants are unlikely to meet all of them, and employers know this. Therefore, don't be discouraged if you do not meet every single requirement, especially those designated as 'desirable' rather than 'essential'.

"When I was looking for a job following my PhD I signed up with recruitment companies to help with my job search. First of all I only considered jobs where I could meet 100% of their criteria, but I soon realised that this was too restrictive as a job-seeking strategy. If you think you can do the job and like it, apply for it even if you don't match all the specifications. The chances are none of the candidates who have applied for the post fit all of them either".
Dr Delphine Oddon, production manager, pharmaceutical company.

---

process for employers by sometimes sifting application forms according to keywords, this rule is taken to the extreme; if you fail to include the right keywords in your application, you are likely to be filtered out during the initial screening process. It is normally the employer who is in the stronger position so it is up to you, the applicant, to get their attention and gain their confidence. You need to convince them that you are a realistic contender who will ultimately succeed in their organisation and achieve results.

Presenting a professional image

Until a few years ago, hard-copy documents were used to make applications. Information was relayed on paper complete with correct layout, indentation and formal address. The application was then posted to the employer in plenty of time to allow for the 'snail mail' system. Nowadays, things have changed radically and email, or some form of electronic submission system, has taken over as the preferred method of delivery. The informal nature of email has given rise to a number of pitfalls which applicants can stumble into if they are not careful. Applications should always have a formal tone and follow most of the previous rules associated with hard-copy applications. Addressing the employer too casually, typing errors and sloppy presentation can all engender a bad impression, leading to rejection.

In addition to using accurate wording and matching yourself to the specifications, ensure your grammar, spelling and layout are faultless, as this creates a good impression and reflects your commitment to the job/course for which you are applying. Don't rely on 'spell-check'. Get a friend or colleague to read your application to double-check for avoidable errors. Always consider the person on the receiving end of your application and market yourself to them. There are usually exceptions to any rule but general guidelines exist which you should adhere to when completing application forms or compiling your CV; you can always adapt them when you feel confident to do so.

> **Box 6.2 CV or resumé**
>
> What is the difference between a CV and a resumé? In many countries, the terms 'CV' and 'resumé' are completely interchangeable and mean the same thing: they are documents which are used to relate information about a person's employment history, education and other relevant experience. Both documents serve the same purpose and are used, primarily, to apply for jobs.
>
> - *Resumé*: in countries such as Canada, USA and some in Europe, the word resumé is more commonly used. This is a 1–2-page document, whereas the CV is longer and can be anything up to 20 pages. Here, the CV includes a comprehensive account of a person's career history.
> - *CV*: in other countries (such as the UK), the term CV is more commonly used and is a two-page document (unless one page is specified). Extra pages can be appended to a CV in order to include additional information, such as a list of publications.
>
> In this book, for the purposes of consistency, the term 'CV' is used and refers to a two-page document, as specified above.

Methods of application

There are many circumstances in which you will need to use an application document: (1) responding to a job advertisement; (2) applying for a fellowship, funding or membership; (3) making speculative enquiries; and (4) attending a careers fair. However, there are only two fundamental methods by which you can make an application (Box 6.2):

- *curriculum vitae* (CV) – the most common form of application
- an *application form* – primarily used when making applications to large organisations and educational institutions.

Curriculum vitae

The primary function of a CV, also known as a personal history or resumé (see Box 6.2), is to get you to the interview stage. Sometimes an employer may state 'apply in writing', which means they are asking you to send your CV with a covering letter. Your CV is your personal advertisement 'selling' you to someone who has most likely never met you. This is quite a challenging task and much harder than a face-to-face meeting. How do you get their attention and interest without the advantage of body language, eye contact and tone of voice? How do you stand out from the crowd and persuade the employer to consider you for interview? Your CV is a highly dynamic and flexible document whose aim is to promote you at your very best and to get you noticed. The following section sets out guidelines to help you to write an effective and high-functioning CV. Examples of six CVs and an analysis of how they have been adapted to six different job specifications are illustrated in Appendix 3.

Structure
Imagine how you might choose to buy a newspaper. First, you look at the front page and your attention is drawn to the most bold and eye-catching typeface,

words or image. If this takes your interest, you scan the page to see if you want to read more and, if so, you may decide to buy the paper to see more detail on the inside pages. Newspaper editors do this to sell their daily publications in the face of extreme competition from others. They market themselves to their specific audience by using particular words, layout and images, a strategy which would not work for another type of audience, who will be drawn to another newspaper genre.

The strategy you use for your CV is not far removed from that of the newspaper editor. Its layout, structure and words will be attractive to one employer but not another, depending on what they are seeking from their applicants. They will have an image in their head of their ideal candidate, clues to which are evident in their job specification. It is the task of your CV to sell you in the face of what is likely to be quite considerable competition. You need to catch the eye of the reader first who will then scan your CV and, if s/he likes what s/he sees, will spend time reading for more detail. If your CV technique is good, you will ensure that its structure highlights all the most significant facts about you relevant to the job description. It will convey your enthusiasm and commitment to the post as well as some insights into the organisation. Therefore, the way in which you structure your CV, the information you include and the vocabulary that you use will be crucial considerations when compiling your CV (see Box 6.3 for examples of strong action words).

As a general rule, you should always put the most relevant facts on the first page, and as near to the top of the page as possible. This involves moving your information around and reorganising it according to each job you apply for. Having a basic CV (see page 72) is fine; it can be used to apply for jobs very closely matched to

---

Box 6.3 Action words

Use strong adjectives and verbs to express yourself in your applications. Examples of positive verbs (used in the present or past tense, depending on the context) to describe tasks and achievements are listed below.

- Supervise/co-supervise (e.g. two Master's students)
- Co-ordinate (e.g. the activities of the departmental postdoctoral association)
- Manage/co-manage (e.g. a complex research project within a team of four)
- Lead/take a lead role (e.g. in organising the faculty seminar series)
- Secured funding (e.g. with my PhD supervisor to extend the project for a further 3 years)
- Developed (e.g. a protocol to investigate xxx resulting in xxx)
- Investigated (e.g. avenues for further funding which led to a successful bid …)
- Organised (e.g. an event for our student social group …)
- Prepared (e.g. a report on my research findings for the funders and other collaborators)
- Presented (e.g. a poster at an international conference)
- Initiated (e.g. a link with a postdoc in another research group leading to a new collaboration)
- Take responsibility (e.g. for keeping our research group informed about progress)
- Analyse (e.g. and interpret complex data …)
- Completed (e.g. a major research project on time …)
- Created (e.g. a more efficient ordering system within the lab)
- Demonstrated (e.g. my work to groups of school students during an open day)

your current post, or act as a working document to be adapted for other vacancy applications. However, if you use the identical CV again and again for every job application it will not be performing its purpose to best effect. The employer may notice this and overlook your CV in favour of more targeted applications.

Length

Although the literal translation of *curriculum vitae* is 'course of life' you should only include the most relevant information in your CV rather than your full career history (Box 6.4). To keep your reader's attention, the main body of your CV should only be two pages long so you will need to exclude extraneous information to achieve this. Basic rules such as good use of space and layout, using bullet points, headings and short punchy sentences will ensure the information has impact. If you need to include lots of publications, you can list one or two key papers in the main body of the CV and then provide a full list using appendices.

Only submit a one-page CV when this is specified as you will probably be underselling yourself otherwise (although this rule may differ between countries – see Box 6.1). One-page CVs are the hardest to write and you will need to be extremely discerning about the information you include. To help save space, it is normal practice to omit referees and to state 'Referees can be supplied on request'. This tactic may also be used if you would rather your referees not be approached unless you receive a definite job offer.

Format

You should structure your CV into headings using short sentences and bullet points. Make the best use of space and ensure the typeface is consistent. Use of bold, underlining and indenting draws the eye down the page. Don't be tempted to use colour unless you are applying for a more unconventional

---

Box 6.4 Common CV mistakes

- *Too long*: unless you have reached a very senior level, two pages are normally adequate for your CV (you can add extra information such as publications using appendices).
- *Untargeted/generic*: you always need to match your CV to the job description.
- *Spelling mistakes, poor grammar*: there is no excuse for these errors, which may be the reason why your CV is rejected by an employer.
- *Disorganised*: the information should be laid out logically and consistently.
- *Too many irrelevancies*: don't include irrelevant information – it gives the impression you are not informed about the job and haven't taken the time to target your application.
- *Too sparse*: don't undersell yourself. Make sure you include all relevant information and avoid gaps in your experience.
- *Overwritten*: too much information in a small font with narrow page margins can be overwhelming and difficult to read.
- *Unconventional*: you may be tempted to present your CV in a more colourful and eye-catching way. Only do this if you are confident it will work in your favour, otherwise employers may play it safe and avoid you.
- *No covering letter*: as a rule, all CVs should be accompanied by a one-page covering letter setting out why you are interested in this job and the key skills and experience you possess which match the job requirements.

creative job – stick to black on white as a rule. With the advent of social media, blogs and personal websites, it is now possible to extend your CV by providing links to social media sites so employers can see more extensive information about you.

Space allocation and the order of appearance of particular sections on your CV will highlight their importance. For example, your PhD should be positioned first in the 'Education' section and be given more space and contain more detailed information than other less advanced qualifications (unless they are more relevant to the application).

The most important question you need to ask yourself is 'Am I highlighting my key selling points?'. You can test this once you have completed your CV by showing it to a friend or colleague for 30 seconds and then asking them to tell you which information they can recall. This is probably what the employer will see on an initial scan of your CV. If they notice something of interest, they are likely to read for detail. If they like what they see, you are well on your way to being selected for interview.

Types of CV

A CV should be tailored according to its purpose and will need to be adapted, even if only very marginally, each time it is used. To this end, you will need to shape the structure of your CV according to its function. There are three main types of CV (examples of which can be viewed and compared in Appendix 3): chronological, targeted and skills based.

Chronological

The chronological CV is used when making applications for jobs which are closely aligned with your current post, e.g. applying for a postdoctoral position where the research and skills required are very similar to what you are doing now (see Appendix 3, CVs 1 and 5). It should, more accurately, be named the 'reverse chronological' CV since it conveys your career history in date order beginning with the present and working backwards to previous experiences. The chronological/basic CV can also act as a working document from which targeted CVs are formulated for making a range of applications.

Targeted

The targeted CV is the most commonly used version. Here, you rearrange your information into sections in order to highlight the most significant experience and skills so that your CV is targeted as closely as possible to the job description. A reverse chronological CV can easily be converted into a targeted CV to highlight particular knowledge, skills and experience relevant to the post for which you are applying. For example, for a research post in industry where advanced knowledge and experience of certain techniques are needed, instead of spreading your evidence amongst two or three postdoctoral posts, you can bring it all together into a separate section entitled 'Research and technical experience'. For a science communication post, you could consolidate disparate activities carried out during your PhD and postdoctoral research into a section entitled 'Science communication experience'. If teaching is required and your experience is hidden within a post you held many years ago, you can bring it to the fore by introducing a 'Teaching' section and including any other associated experience. For examples of targeted CVs, see CVs 2, 3 and 6 in Appendix 3.

Box 6.5 What to include in your CV

- *Personal details*: contact details and social media links
- *Career goal/key capabilities*: only include if it has impact and adds significantly to your CV, otherwise incorporate into your covering letter instead.
- *Employment history*: highlight the most important posts with associated skills and experiences which match the job specification.
- *Education*: list in reverse chronological order, giving your PhD or highest qualification most prominence by placing it first. Provide a précis of your research project according to its importance to the application.
- *Skills*: skills and personal qualities may be listed separately or incorporated into your work history section, depending on how you are targeting your application.
- *Other information*: use this section to include additional information which does not fit into other sections, e.g. achievements, awards, membership of a professional body or society, patents, funding, etc. Again, this information may be better placed incorporated into other sections.
- *Interests*: including an interests section is debatable. In some countries interests should definitely be left out but in others, such as the UK and US, employers like to see a more personal dimension to a CV, which can give further clues about whether you will be suited to the role or company.
- *Referees*: usually your immediate and previous supervisor, employer or tutor; referees are supposed to verify your abilities and recommend you.

Skills based

If you are aiming for a new career such as administration, finance or general management, a skills-based CV will be the most appropriate. A skills-based CV is a useful format for applications where you have no directly relevant experience. Here, you will need to demonstrate the transferability of your current skills and experience into a new career. You can group together activities which provide evidence of the same skills specified in the job description and title these sections, for example, 'Communication', 'Team working' and 'Project management' skills. Evidence of personal qualities such as integrity and self-motivation adds weight to your application. See Appendix 3, CV 4, for an example of a skills-based CV.

Content

General rules about CV content are described in this section (Box 6.5). However, conventions exist in different countries which may differ slightly from the advice given here, so you will need to refer to other more specialist information to check against the rules set out there. Such information can be found, for example, at www.iagora.com/iwork/resumes/index.html and www.prospects.ac.uk/country_profiles.htm.

Personal details

You need not title your document 'CURRICULUM VITAE' – use your name as the header for your 'advertisement'. Your contact details then follow underneath (although some people prefer to put them at the end of their CV and this is acceptable). Include contact details such as your address, email, phone number (e.g. work, home and/or mobile/cell number) and, if you have an address, your Skype name. If you are on social media such as Twitter, Google+, Facebook and LinkedIn, or if you have a personal website or write a blog, include this information too (see Chapter 5 and Appendix 2 for more on social media). Employers are likely to look you up (Macfarlane 2012) so only include links to your professional or semi-professional profiles which reflect your talents, knowledge and personal attributes. You want your links to enhance your image, not damage it, so omit anything too personal and check privacy settings.

Other information can be included at your discretion (or if specifically requested). For example, gender, date of birth, nationality, passport number and marital status. In some countries it is convention to include a photograph at the top of your CV. However, unless this is the case or specified in the application guidelines, it is generally not necessary to include it.

Career goal/intention

Many CVs carry a short statement at the start displaying the person's career goal (see Appendix 3, CVs 1, 2 and 4). I would warn against such statements unless they are written in a highly targeted and effective manner, otherwise they may look shallow and clichéd. If you want to include some key facts linking you to the job role, you can list/bullet-mark three or four *key capabilities* (see Appendix 3, CVs 5 and 6). To do this most effectively, look at the job description and highlight the essential skills and qualifications required for the role. If you can fulfil all these requirements, list your evidence very briefly at the top of your CV. You can then reiterate them in the body of your CV along with a brief description to extend the information. Key capabilities should also be highlighted in your covering letter (see below) so including them in your CV may be wasting valuable space.

Education and qualifications

Some people separate their educational institution from their qualifications. However, this can be confusing. Position them together in one section and list them in reverse chronological order starting with your PhD and working backwards to your Master's, undergraduate degree and then (optionally) your school/college leaving qualifications. Provide the start/finish dates of each qualification and the name of the institution.

The amount of information you include about your PhD and other university qualifications will depend on the level of academic attainment required for the job.

- *Academic posts*: provide the full title of your PhD and a short abstract, main aims and objectives. You can also include key techniques or skills if they are essential to this post. If you have a long list of publications, include one or two key papers here and add an appendix for the complete list. Follow this on with your Master's and undergraduate degrees with similar levels of specificity but reduce the amount of information as this is generally less important than your PhD.
- *Non-academic posts*: rewrite the title of your PhD and provide a brief description using language which will be understood by the employer. More general scientific terminology is appropriate for a research job in industry and, for non-scientific jobs, simplify the description to the level of public engagement.

If you have relevant vocational qualifications or have attended short courses and occasional workshops, list them in the education section after your academic record. If one or more of these additional courses is of particular relevance to your application, rearrange the list so that they appear first and can be seen more overtly. Alternatively, include them later in your CV under 'Additional information' or 'Achievements', depending on their importance to the post.

Employment history

As a general guideline, the first section of your CV should relate what you are doing now. If you are a postdoctoral researcher, fellow, lecturer, assistant professor or working in industry (rather than undertaking a PhD), it is convention to place your employment history first on your CV, ahead of your education section. In those countries where doctoral students are regarded as professional staff, PhD information can be placed in the employment section. Follow on from your current post with previous work experience, which may include previous research posts and non-related experience.

The amount and depth of information you include will depend on the post. You can bring together relevant work experience by subsectioning the CV into 'Relevant work experience' (e.g. 'Research experience', 'Management experience') and 'Other work experience'. Relevant work experience could include voluntary placements or internships which may be the most important evidence of your suitability for the job.

- *Academic posts*: the title and aims of the project, short abstract and other key information should reflect the essential requirements of the role. Subsection essential outputs such as grant raising, publications and collaboration and include a brief punchy description for each. As with your PhD, if you have a long list of publications and conference presentations associated with your research, follow the advice above and make use of an appendix.
- *Non-academic posts*: following the rules from the 'Education' section above, relate your research experience in terms which will be understood by the reader of your CV. Dilute down the science according to the requirements of the job and translate it into more general skills (see Box 6.6). You can reorder your employment history by including subsections which group together relevant experience.

Skills

Many researchers define themselves in terms of their specific knowledge and research interests. However, if you examine the vacancies listed in Chapter 4, you will notice that the majority of employers specify skills and personal qualities. Therefore, skills are a highly important part of your CV and should be presented as coherently as possible. Rather than spreading them out amongst the various education and employment sections on your CV, you can bring them together into a single section so that you consolidate your evidence.

The 'Skills' section should not only contain technical and research-related skills, it should also demonstrate evidence of your personal skills, depending on the job specification. If 'management' skills are stipulated, then include this as a subdivision in the Skills section and bullet point a list of evidence which demonstrates your management experience. If the advertisement is asking for excellent 'communication' skills you can use the same evidence but reword it so that it conveys your communication abilities.

The 'Skills' section is even more important in a skills-based CV (see page 73) as it brings together the transferable skills gained during your employment, education and other activities into one place on your CV. In this case, you can even place the 'Skills' section first on your CV immediately below your personal details and before your 'Education' or 'Employment' sections.

A separate 'Skills' section is not always necessary and the information is sometimes better placed elsewhere such as within the 'Employment' section. Keep in

**Box 6.6 Providing evidence of transferable skills from your
PhD/ postdoctoral research**

Your research experience has given you a vast range of skills to promote to a new
employer. The following (abridged) job advertisement may or may not be of interest to
you. However, every PhD-qualified researcher should be able to provide evidence of
the requirements specified.

Example job vacancy: Project Manager (university department)
Aim of the role
To project manage and implement new initiatives within the School of Social
Sciences under the direction of the Head of School. Projects already identified include
the introduction of a new Master's scheme, the organisation of focus groups and
enhancing international student recruitment. The post holder will liaise with academic
staff as well as staff in a variety of the University's support services and with representatives of external organisations.

Applicant requirements
- Evidence of the ability to **convey information clearly and accurately in a variety
  of formats, explaining complex information** in a way that is **understandable to the
  target audience**.
- Evidence of the ability to **collate and analyse data** from a variety of sources, drawing
  relevant conclusions and identifying options for future action.
- Evidence of the ability to **organise, prioritise and plan time and resources**, including
  personal **time management, planning work for others, operational planning and
  strategic planning**.
- Evidence of the ability to **develop networks** of useful contacts within and beyond the
  Centre, the School and the University; influencing developments through personal
  networks.
- Ability to deliver a high standard of **project management** service to the School,
  promoting to others, and setting and maintaining overall service standards.
- Ability to **identify and develop options and use initiative** to select solutions to prob-
  lems which occur in the role.
- A **flexible and adaptable** approach to work.
- Willingness to **develop new skills** and **take on new challenges**.

Table 6.1 Evidence matched to the job specifications for the Project Manager
vacancy listed in Box 6.6.

This table demonstrates how your research experience can provide evidence for
the many and varied skills and attributes set out in this job advert. Why not add
in your own and extend the list to see how well you match up?

Job requirements	Evidence
Convey information clearly and accurately in a variety of formats. Collate and analyse data. Explain complex information so it is understandable to the target audience	Generate complex datasets derived from multiple experiments which require clear and accurate communication. Able to select appropriate modes of delivery using charts, tables and graphical diagrams. Regularly produce short reports and disseminate results to my research group. Have presented papers and posters at national and international conferences.

Develop networks and contacts	Use social media and networks extensively in order to share experiences, information and latest developments with other researchers. Set up Twitter hashtags for specific projects and events in order to draw in like-minded professionals. Member of LinkedIn groups to which I regularly contribute. Network face to face at meetings and conferences which has yielded many new contacts and led to a new collaboration with a German research group. Member of the department's postdoctoral association which involves attending events and networking with other staff in the Faculty.
Project management Organise, prioritise and plan time and resources, including personal time management, planning work for others, operational planning and strategic planning	Project manage the planning and implementation of my current research project following on from my PhD. Able to prioritise tasks balancing experimental work with data analysis and communication. Assist in the co-ordination of our research collaboration which comprises two departmental partners and researchers in three European countries. Adhere to strict milestones set out by the European funding agency, contributing to update reports. Highly self-motivated and always meet deadlines on time or ahead of time.
Working with others	Collaborate with colleagues of different disciplines within my research project to co-ordinate activities and share data. Regularly in contact by email with European and industrial partners and have made two visits to Germany to work more extensively on specific areas of the project.
Flexible and adaptable approach to work. Willingness to develop new skills and take on new challenges	Unexpected results from a recent project required me to learn a new technique and reorganise my resources.

mind that your CV is a fluid and dynamic document and information can be moved around in whichever way presents you at your best.

You will have your own set of experiences and evidence which is personal to you. You may also be able to provide further examples from non-work-related activities. If you have evidence to demonstrate that you exceed expectations or you have something extra to make you stand out, it will help to enhance your profile further.

Other information
You may have important additional information that you would like to include in your CV but there is no obvious place for it. In this case, you can create a new section and title it depending on the theme of the information to be included.

If it relates to medals and awards, positions of responsibility or membership of a group, you could title the section 'Achievements and interests' or 'Responsibilities and achievements'. Perhaps you have attended professional development courses which are worth listing or you may have accumulated commercial experience, such as collaborating with industry, development of a patent or participating in an entrepreneurship scheme. Whatever your experiences, consider carefully whether you need an additional section in your CV and ensure that the information is relevant to your application.

Interests

This section is slightly controversial and its inclusion is country dependent. In the UK it is normal to have a brief 'Interests' section which adds a personal dimension to your CV. Outside pursuits such as social clubs, sport, travel and music reveal something more about your character and personality, which many employers find useful during the selection process. However, in many other countries this information is viewed with suspicion – why would they want to know about your social life? Therefore, you will need to use your own discretion about this section. A notable exception to this general rule is if you are changing career direction and your interests are central to the new role. In this case, you can introduce a section at the start of a targeted CV (see page 72) entitled 'Relevant experience'. This information will be far more important than your research and education, which may be better placed on page 2 with little or no mention of your publications (see Appendix 3, CV 3 for an example).

Referees

The normal procedure for listing referees is to provide the name and contact details of two people (or more if required) who can give authoritative (and, ideally, differing) accounts of your personal and professional qualities. Normally, use your most recent and previous employers/supervisor. You can provide a personal contact if they are in a position of authority (but not a family relation). If possible, make sure your referees know and understand the requirements of the job so that they can promote you in the best way. Don't be afraid to coach them on the most important aspects of your experience which you would like them to highlight according to the requirements of your new post. Ideally, your referee should be your supporter and want to promote you at your best so send them a copy of your CV and brief them on the job you are applying for.

Publications and presentations

If you have one or two recent or high-impact publications you would like to highlight, position them in the PhD or postdoctoral research section of your CV. If you have a long list of papers it is useful to add an appendix so that they do not break up the flow of your CV or interrupt other sections. An appendix can be as long as you like as it does not interfere with the flow of the main document. Include journal papers (published, submitted and in preparation), presentations made at conferences or to other audiences, grants, patents and other research-associated outputs. However, be cautious about the amount of information you list if you are not applying for an academic research position. A lengthy list of papers and presentations is irrelevant to some applications and will be of no interest to the employer. It may also give the impression that you are still tied to your research, or are leaving it unwillingly.

Normally you should list your publications in reverse chronological order using a consistent citation style with your own name in bold. If you have one or two publications you would like to single out for special attention, highlight them by introducing a subsection at the start of your list entitled 'Key publications' and then follow it on with 'Other publications' (published and 'in press'). Include submitted and 'in preparation' papers in a separate subsection (try to get them written as soon as possible). If you have contributed articles to general interest magazines or if your work has received media coverage, this is also evidence of communicating your research. In this case you can include further subsections entitled, for example, 'Public interest publications' or 'Media exposure'. Following your publication list, you may wish to add a list of papers and posters presented at conferences. Again, highlight particular talks you have presented if you were an invited speaker, for example.

If your experience is limited and you have few publications to list, use as many examples of your communication as you can even if it includes talks you have presented in your department. However, bear in mind that for most fellowships and more senior research posts, your publication record will be critical to your application success so if you have been neglecting to write up your research, make this a priority; it will form a major part of your future career if academic research is what you are planning.

Covering letter

Introductions are an important, and usually formal, procedure when meeting people for the first time. Presenting your CV also requires a formal introduction and this is done using a covering letter. The purpose of the covering letter is to create interest and encourage the reader to look at your CV. You need to match it very precisely with the employer's specifications, highlighting three or four key points from your CV. The information can be reiterated in the 'Career goal' or 'Key capabilities' section (see Box 6.5) of your CV as well as in the main body. When you send your CV via email, you cannot be sure which file will be opened first so repeating yourself in both documents is not a waste of time and will ensure that the important information is seen quickly and easily.

As a rule, covering letters should be one page in length unless you are applying for a senior academic post where you need to present a more in-depth account of your research experience. Even here, you can append a separate document for your research statement so that your covering letter is kept brief and succinct. In addition to your address, date and job reference at the start of your letter, the four main sections/paragraphs to include are as follows.

1 A general introduction: say where and when you saw the job advertised and state briefly why you are interested in the post.
2 Say what you are doing currently and your motivation for applying for the post in more detail with 2–3 pieces of evidence to show your suitability for the role.
3 Demonstrate how your experience and skills will contribute to the company/ project, citing your knowledge of their business/research and demonstrating your suitability.
4 Finish on a positive note, saying you hope you will be considered for the post and look forward to hearing a decision.

When writing to a named person, the convention is to sign your name 'Yours sincerely'. For 'Dear sir or madam', use 'Yours faithfully'. See Appendix 3 for an example of a covering letter.

Writing your CV

Writing your CV from scratch will probably take longer than you anticipate so make sure you start soon! This is especially important for those who are actively job seeking. If you see an unexpected job opportunity, have a chance meeting with a group leader or potential new employer who invites you to send them your CV, you will be glad you already have a basic document drafted. Start with the chronological CV (see page 72) from which you can extract relevant information for your working CVs. Depending on how you normally prefer to do things (in a linear way or more scattered), you can take a number of approaches to start gathering together the content of your CV.

- Make a list of all your personal information and experience to date (you may wish to start with general headings and then add in details to each).
- Alternatively, mind-map your information on a large piece of paper or onto sticky notes (pick out major headings and link them with associated information).
- If you already have a personal development plan or have carried out a skills audit (see Chapter 3), you will be able to draw upon this information and transfer it directly into your CV.

The object of this exercise is to document everything relevant that you have done to date. This includes employment, education, training, outside interests, voluntary work, awards/achievements and so on. It is easy to forget important information, such as conferences or courses you have attended, involvement in events from a few years ago (but do not go back too far) or previous achievements which may be relevant to the application. If you are a mature PhD student or postdoctoral researcher, make sure you include previous work experience. Once you have all your content recorded, you can start to summarise and organise it into sections, as set out in the guidelines above.

A word of warning: many researchers are concerned that employers in non-science sectors will consider them to be overqualified and question whether they should remove information about their PhD from their application and minimise their experience. I would advise against this (refer to career narratives 18 and 19 in Appendix 1 for additional confirmation). There are ways in which you can present yourself to an employer to highlight your career history so that it is relevant to the job. You will need to justify your transition into this career but, with careful thought and positive wording, your application should demonstrate your genuine enthusiasm for changing careers. A good understanding and knowledge of the job market (Chapter 4) coupled with self-awareness (Chapter 3) should optimise your chances of success.

Summary

These general guidelines aim to help you to craft effective CVs and covering letters which are targeted to achieve their purpose (see the examples in Appendix 3). CV 'wizards' and other formulaic documents are available to download from the internet which can act as a useful starting point, but use them with caution as they are not ideally structured for you as researchers. You may find it difficult to present yourself at your best and, furthermore, employers become used to seeing these standard CVs and it can give a negative impression. You may have heard of the Europass CV, which has been introduced to assist and encourage people to write their CVs and to present themselves effectively (available from: www.uknec.

org.uk). You could use this as a template for storing your CV information, as in a personal development plan. Some European employers stipulate that applicants apply using the Europass CV so I will leave it to your discretion whether to use it. However, as a general rule, taking the time to personalise your CV using your own format and structure and presenting yourself as an individual will always reap the best results.

Application forms

Application forms are covered more briefly here than CVs. The content of a CV is similar to that of an application form, except that the information presented in an application form is under the control of the employer. The way in which you convey your experiences and skills is steered by the structure of the application form and questions asked by the employer. Sometimes the format of an application form can be frustrating for applicants, as the standardised design used across the company does not always contain relevant questions for posts in every department. However, the sections normally include:

- name, personal details
- education and training
- employment history
- specific situational questions
- a large blank page asking why you are applying for the position or research grant, including your experience, skills and attributes (with examples) to support your application
- interests/achievements/awards/affiliations
- referees.

Most sections are straightforward, usually requiring you to list your education, qualifications, previous and current employment history with dates and a brief description of your main duties and responsibilities. In these sections it is important to include all your personal information, ensuring there are no unexplained gaps in your career. For example, if you took maternity leave or were unemployed, this should be accounted for. Relevant voluntary, temporary or part-time work should be included and other experience may demonstrate evidence of personal qualities, such as flexibility and self-determination. Unless the application specifically states otherwise, group particular areas of work into sections so you provide an encompassing description with evidence of associated experience and skills.

You may have some anomalous information which does not fit into the structure of the application form. In these situations use your judgement or seek advice from a friend or colleague as it is probably not crucial to the success of your application. Don't dwell on any negative (as you see it) information. This could be a period of unemployment, a failed exam or low grade, or perhaps a low number of publications compared to your peer group. Always be upbeat and positive and do not insist on making excuses or drawing attention to them – no-one is perfect! It is common for people to target the (usually) very minor 'black marks' on their copybook. Consider all the overriding good points you have to make and, although in an application form you cannot get away from citing the 'bad' points, give them cursory mention and be prepared to talk about them at interview.

Making applications

Ch 6
Making applications

Specific questions/evidence

In some applications there are sections which ask you specific questions related to the post. Provide detailed evidence (one or two examples) to back up the statements you make in these sections. Some questions are situational, e.g. 'Give an example of when you worked in a group. What was your role? How do you negotiate with others? How do you deal with failure? Which personal attributes do you possess which make you suited to this post?'. You will be questioned on your answers at interview, so make sure you revise your application form and are well prepared. Do not be tempted to exaggerate (or lie) on your application form. If you cannot answer the questions, it may be an indication that you are not suited to the job or that you need to gain further experience before putting in an application.

Broad questions/evidence

Many application forms, instead of consisting of specific sections, will provide one large blank space ('continue onto an extra page if necessary') inviting you to explain why you are applying for the job, grant or course. You will need to structure this section as you would an essay or report so that it is paragraphed, with subheadings and bullet points referring back to the specifications of the job description or the funding criteria. Don't be afraid to mimic exactly the wording which the employer has used in the specifications – use this for your subheadings and in the text of your application as this will catch the attention of the employer. This is crucial to the success of your application. If you have a CV with an up-to-date account of your experience and skills, you can draw on your personal information for this section of your application.

Conclusion

An application is much like writing an essay or sitting an exam. You need to read the job requirements and specifications carefully and then tailor your text accordingly. This involves structuring your information and providing evidence of your suitability. As with an exam, you will score most of your points early on, so be strategic and organise your information so that the most important facts appear at the start.

Selling yourself on 'paper' is challenging without the additional armoury of body language which you have in a face-to-face interview. You need to use strong and positive vocabulary to engender confidence in your ability so do not be afraid of 'boasting' on paper – it will portray you as enthusiastic and committed (see Box 6.3). If successful, your application will undoubtedly lead you to the next stage in the transition process – the interview – so make sure your information is accurate and that you are going to be able to talk about it confidently, as if you were defending your PhD thesis.

Reference

Macfarlane M. (2012) *How much personal information should job seekers share online?* Guardian Careers. Available from: http://bit.ly/Ao8faM

CHAPTER 7

Successful interview technique

If you are asked to attend an interview you should congratulate yourself. You are now closer to securing the job! Along with making applications (Chapter 6), interviews are a major part of the transition process as featured in the DOTS model (see Chapter 2). Employers rarely accept people on the basis of their application alone (although some research fellowships are decided solely by application). You have shown the employer/funder/professor that you meet the necessary criteria and demonstrated your potential. Whether it is your research profile, experience, transferable skills or other qualities which brought you to this stage of the process, you will now need to relate this information in a structured conversation with the employer (sometimes in conjunction with other assessments) to prove that you are the best candidate for the job.

It is useful to think of interviews in similar terms to your PhD defence/viva. Although a daunting prospect, they are necessary in order to ensure you are as good face to face as you say you are in writing (Box 7.1). The employer needs to find out whether you are able to answer his/her questions confidently and knowledgably. As well as further exploring the relevant experience you possess and your potential to grow in the role, your understanding of the post and the duties and responsibilities associated with it needs to be checked. Furthermore, the employer wants to meet you to see if you will fit into the work environment and get on with the other team members. Equally, the interview gives you the opportunity to see if the job, the company culture and people will suit you. As with a PhD defence, interviews are not there to catch you out, to make a fool of you or to trap you (or they shouldn't be!). If the employer is to evaluate the interview candidates most effectively, s/he should make the process as transparent and fair as possible so that you have the opportunity to present yourself at your best. In order to do this you will need to be well prepared.

Types of interviews

One-to-one interviews

One-to-one interviews are quite common in academia, especially for PhD or junior postdoctoral positions. They are the only step in the interview process and are usually more informal and so less stressful than other types of interviews. However, be careful not to become too relaxed and forget the purpose of the interview during these more informal encounters. Keep in mind that you are there

Career Planning for Research Bioscientists, First Edition. Sarah Blackford.
© 2013 Sarah Blackford. Published 2013 by Blackwell Publishing Ltd.

Box 7.1 The purpose of interviews

For the interviewer

- To ensure you meet the criteria for the job
- To fill in extra details and extend the information from your application
- To test you further (e.g. presentations, psychometric assessments)
- To determine how well you can communicate and present yourself face to face
- To find out more about your personality
- To ensure you will fit into the organisation

For the interviewee

- To make a good impression
- To sell yourself, your skills and experience to the best of your ability
- To find out more about the organisation and determine whether it is right for you

---

to promote yourself and convince the interviewer of your abilities, even if it is your current supervisor. In some cases you may have to steer the conversation so that you manage to include important information and evidence of your suitability for the post. For example, you may have relevant work experience or are knowledgeable about a particular technique which has not yet been discussed – make sure you get this into the interview discussion.

Telephone interviews

One-to-one interviews are also used as an initial screener for non-academic interviews where the human resources department or a recruitment company has been tasked with vetting the applicants to ensure they generally fit the company's employee profile. Increasingly these are being conducted by telephone. This type of initial interview is becoming more commonly used by companies who may be faced with a multitude of applications. Telephone interviews can feel strange and uncomfortable without the advantage of a more formal one-to-one interview setting and the ability to use body language and eye contact to increase rapport and enhance your impression.

Find out the exact time the interviewer will call you (experienced interviewers will do this anyway) so that you are prepared. You can ask them how long the interview is likely to be and if there is anything you need to prepare for it. If they call you unexpectedly, make your excuses; say you are in the middle of something and could they call you back. This gives you a chance to get your thoughts together and find your application form or CV so you have it to hand during the interview. It may be off-putting and a little strange to answer questions over the phone but be sure that you keep your answers succinct and to the point. There is no eye contact so you will need to use your voice to get across your personality. Answer as calmly and coherently as possible. This is where preparation really counts!

Skype interviews

Conducting interviews via Skype is becoming more common due to more people becoming familiar with the technology and improvements in communications. The use of Skype allows international and national interviewing to take place very cheaply or free of charge. Some researchers have also had to make a presentation over Skype to a panel of interviewers. For advice on how best to handle Skype interviews see www.jobs.ac.uk/careers-advice/interview-tips/1252/job-interviews-by-skype/.

Panel interviews

Panel interviews, comprising two or more interviewers, are common for more senior roles and non-academic positions. Each member of the panel is usually assigned a specific line of questioning so that all relevant areas are covered during the interview. For senior academic posts, the interview panel will most likely be made up of senior staff in the department together with staff from other departments or external collaborators. They will question you on your research profile, e.g. publications, funding success, research ideas, collaborations and your potential to link with other members of the faculty and university/institute.

A similar structure is used for panel interviews conducted outside academia, except that the focus of the questioning will be different: one member of the panel may ask about the technical side of the job, another may question you on your achievements, transferable skills or training. A human resources representative may ask you more general questions to test your personality (e.g. what do you consider to be your strengths? How would your friends describe you?). When responding to questions, direct your answer for the majority of the time at the person who posed it, but occasionally glance across at other members of the panel to make them feel included. If they happen to be looking down at their notes don't be put off by this – they will still be listening. They may simply be studying their next question or rating your performance using an interview proforma.

Presentations

Interviews for more senior roles often include a presentation, depending on the position. Academic posts may require a research presentation or a mock lecture to the whole department. Postdoctoral roles and other non-academic positions are more likely to ask for a short presentation to the interview panel.

If you are invited to give a presentation as part of the interview process, the prospective employer should specify the focus of the talk and your audience (and if they don't, you can contact them to ask for more information). This is important as the content of your presentation and the way in which you pitch it will very much depend on your audience. For instance, if it is aimed at the interview panel you will not need to take account of any students being present (unless they have specified that you are role-playing a lecture). If it is in a lecture theatre to the whole department, you will need to adapt your presentation so that it is broad and inclusive of non-specialists.

Beware of including too much content and try to restrict your slides as much as possible. Make sure you practise your presentation in order to get the timing right, ideally to a friendly colleague who will give you constructive feedback to help you to perfect your performance. Aim to finish 2–3 minutes before the time allotted so that you can invite questions. The worst thing you could do is run on too long as this will reflect badly on you and your time management. They may even cut you off as they need to keep their interviewing schedule to time.

Interview tasks

During some interviews you may be set a short task to test you on a specific aspect of the job which is of central importance. For example, if excellent organisational skills are required, you may be asked to do a 'prioritisation' task, such as an 'in-tray' exercise. Here you will be given lots of information and asked to assess it quickly, prioritise it and use it to make a decision or write a letter, according to the task. During the exercise more information may be passed to you to see how you respond and react. Other tests may include checking your attention to detail (e.g. for an accounting or copy-editing job) or asking you to write a letter to demonstrate, for example, your ability to express yourself diplomatically.

You should be informed of the test when you receive your interview invitation so you are prepared for it (although it is unlikely they will tell you the exact details). Try to stay calm and do not fixate too much on this part of the interview. Employers take account of nerves and the majority of tests usually only count marginally towards the overall interview assessment and final selection. Further sources of information are available to help you to prepare for such occasions (see Appendix 4).

Assessment centres and psychometric testing

As part of the selection process, you are less likely, especially for research and science-related jobs, to be required to attend an assessment centre or undergo psychometric tests. These are mostly associated with business-related graduate management training programmes, fast-track entry routes into large companies, government or public sector positions. If you are invited to attend an assessment centre (for one or sometimes even two days), you will be scrutinised in any number of ways. Exercises will be similar to those mentioned above in the section on interview tasks but with the added dimension of team-orientated activities as well as individual ones. Trained assessors will observe your behaviour in different situations to determine, for example, the way you act in a team, think independently and operate under pressure. They may be looking for leadership skills, creative thinking or self-motivation. If you are invited to be assessed in this way, there are a number of things you can do to prepare for the experience.

- Read up about assessment centres – online information and books (see further information in Appendix 4).
- If your university has a careers service, they usually hold information on assessment centres and may even offer practice sessions.
- Look out for opportunities to take part in assessment centre-type courses or retreats where similar team activities and tasks are in place but in a non-judgemental environment. It will help you to prepare for the 'real thing' but also offers the opportunity for personal reflection and development. Examples include well-established national programmes, such as the UKGrad schools (www.vitae.ac.uk/researchers/15672/GRADschools.html) and events organised by universities during doctoral student introductory weeks.

For any assessment centre or behavioural exercise, try to raise your self-awareness (see Chapter 3) and be yourself (at your best!). The assessors are not looking for people who dominate discussions or who are trying too hard. They want team players who respect others' opinions and are inclusive. Make sure you stand out in the crowd and make considered contributions but stay aware of your situation at all times. You will be being assessed from the moment you arrive till the moment you leave.

Psychometric testing is commonly used by governmental, financial and managerial organisations and departments where numerical, written and comprehension skills are required. You can buy books and sign up for practice tests online which may help you to familiarise yourself with them and improve your performance, but companies are protective of their tests and you won't get sight of the exact test beforehand.

Interview content: what questions will you be asked?

Interview questions depend on the information which the interviewer is trying to find out about the applicants. Specific questioning will be used according to the role and job specifications, while general questions will always feature (see Box 7.2).

Junior postdoctoral interviews

Generally speaking, if you have been studying for a PhD or working as a postdoctoral researcher in academia or in a research institute, you may have had quite limited (or no) experience of structured formal interviews. Professors and other senior researchers tend to prefer a more informal method of interviewing with a one-to-one or perhaps two-to-one set-up focused primarily on the research project. When you were interviewed for your PhD studentship, you were probably questioned about your undergraduate degree, specialist knowledge and practical projects associated with it. There may have been some additional questioning if you have had associated work experience, for example, but the line of questioning and the areas of most interest to the supervisor would have been keenly focused on and around your research potential. Fundamentally, s/he needed to determine whether you were capable of undertaking and completing a PhD.

If you apply for a junior postdoctoral position, it means you are committing yourself further towards a career in academic research (even if this career does not pan out in the long term). You are applying for a salaried position and so will need to work within stricter guidelines according to the aims and objectives of the research project. Therefore, the interview is likely to be more structured towards your research capabilities, technical expertise and dedication to the project. Even though the interview may still be a one-to-one or two-to-one set-up, the line of

---

Box 7.2 General questions asked in any interview

- Why do you want this job?
- What can you offer over and above other people?
- Where do you see yourself in 5 years' time?
- What are your strengths?
- What are your weaknesses?
- What has been your greatest challenge?
- Tell us about your greatest achievement.
- Do you have any questions to ask us?

Box 7.3 Example questions asked during junior postdoctoral interviews

- Tell me/us about your PhD. Why did you do a PhD?
- What did you enjoy most? What didn't work?
- What was the most important discovery you made during your PhD?
- How does that work relate to this current post?
- Why are you interested in this postdoctoral position?
- What techniques are you familiar with?
- Have you identified any gaps in your knowledge/expertise relating to this project? How would you go about addressing these gaps?

---

questioning will differ from that of a PhD interview (Box 7.3). Your primary role as a postdoctoral researcher is to contribute productively to the research project, designing experiments and generating good-quality data which will result in publications and the prospect of further funding.

Thus, you will probably be asked to talk extensively about your PhD research project or other current/previous research work. This will include general and specific questioning about your project, not only in terms of the work you conducted but your thinking behind some of the directions your research took. The interviewer(s) may want to know how much of it involved your input and original thinking so that they can determine your potential to progress in an academic field. Alternatively, or in addition to this, they may want to question you about the challenges you have faced (and how you overcame them), additional associated research activities such as membership of societies or involvement in science communication, supervision and departmental activities. Your publications and research profile will probably be limited at this stage of your career and so your potential to be a successful researcher will be of more interest.

Senior postdoctoral researcher/group leader

Things start to get more serious as you move from a 'junior/middle' to a 'middle/senior' management role. If you are applying for more senior research roles, you are taking a further step along the academic career path. This means you will be judged more seriously on your research profile and associated impact factors such as the number and quality of your publications, your national and, increasingly, international standing, funding record, teaching, supervision and additional administrative and collaborative activities (see Box 7.4). Depending on the nature of the position, your connections with industry, patents derived from your research and examples of innovative thinking will all count towards your success at interview.

The more senior the role, the more evidence of management and leadership will be required. For principal investigator roles, you will need to demonstrate vision and the capacity to lead your research in a direction which will position you and your institution as leaders in your field. For more teaching-orientated roles, evidence of innovative practices and engagement with the pedagogical community will be of primary importance. The panel of interviewers will probably include members of the research group, the department, human resources and even other more distant departments or representatives from industry. You can ask in advance who will be on the panel, which will help you to pitch your answers at the appropriate level.

Box 7.4 Example questions asked during senior researcher/ academic interviews

- Tell us about your research career.
- Where would you look for grant income?
- Where do you see yourself in 5 years' time?
- What research questions do you hope to address in your first 5 years here?
- Given unlimited resources, how would you take this field of research further?
- Summarise your understanding of this field in lay terms.
- Tell me about your teaching philosophy.
- How do you see yourself as adding new energy and direction to the department?
- What do you see in terms of potential for collaborative research with existing members of the department?
- Tell us about your publication record. What are the highlights?

One professor told me that he sends candidates the grant application and asks them at interview to critique it and discuss how they would develop and implement the project. He also asks them how they would drive it differently. His motivation for taking this line of questioning is to test the critical and creative thinking of the senior postdoctoral researcher and their potential, ultimately, to lead a research group.

For group leader posts, broader questioning aims to find out how the candidate would develop the research beyond the horizons of the current project. There needs to be evidence of 'clear blue water' between them and their previous supervisor to demonstrate their research independence. These candidates might be asked to submit a two-page grant application as part of the application process, after which they will be questioned on it at interview. In addition to research-related questions, they may also be quizzed on leadership and management ability by a representative of the human resources department. Example questions include the following.

- Describe a conflict situation you have found yourself in. How did you deal with it?
- Give an example of when you were managed. What do you see as being good and bad management?
- How have you managed laboratory resources?
- What would be your priorities when setting up a new lab?

Non-research posts

If you are moving further afield and away from bioscience research, you may feel you are on less familiar territory when it comes to the interview process. However, bear in mind that you have been selected for interview so you have shown the employer that you have skills and qualities of value to this post. Therefore, you should approach the interview with confidence, knowing that you have already passed the initial selection stage and now need to prove yourself at interview.

Use your application form/CV and the job description to determine what made it successful. If you promoted the transferable skills associated with your research and other experience, be prepared to be questioned on this and use

---

Box 7.5 Example questions asked in non-research post interviews

It is less easy to provide examples here since this section refers to a whole range of jobs and roles. However, some predicted general questions may include the following.

- Why do you want to leave research?
- How does your research experience transfer into this role?
- What attracts you to this profession?
- Why are you interested in working for our company?

There will be specific questioning on the role for which you have applied and, if it is a technical role, you should expect quite detailed probing.

---

evidence to back up your answers (Box 7.5). For job interviews where you have no directly relevant or previous experience, you may be asked what you would do in a particular situation or circumstance. This could be done as a role play or you may have to talk through what you would do/how you would respond/the correct chain of command/who you would include in the decision-making process, etc.

Answering the questions

An interview is a conversation with a purpose and should always be focused on its central objective – to get you the job. Even in more informal situations, keep this aim in mind and ensure you relate your information in an impressive and effective way. Always answer in the context of the job for which you are applying and stay 'on message'.

Revise the job description and keep in mind what the employer is looking for. They are interviewing you to check the information on your application and find out more about you. Imagine you are bullet-pointing your answers and relay your responses in a structured and coherent way. Use real examples supported with details to bring your evidence to life and make it sound solid and convincing. Open questions such as 'Tell me about your research' are equivalent to the blank spaces on application forms which ask you to provide a précis of your experiences. The same rules apply at interview: keep it structured, relevant and concise. If you start to stray away from the point, the interviewer will wonder what the significance of the information is to them.

Do not go into great detail about your research project in a non-academic interview. Instead, speak in more general terms relating your research experience to the job role. Demonstrate, using examples, how you will be able to fit in with the project and help it to achieve its aims. For interdisciplinary research roles, tell the interviewer how your research work and skills will complement and enhance those of the other group members.

Always use upbeat positive language coupled with similar body language to convey self-confidence (even if you don't feel it!). It is surprising how many people who 'look good on paper' disappoint employers by giving a poor performance at interview. Interviewers want you to do well so they can fill their vacancy. Believe in yourself and your abilities and the employer will believe in you and, with any luck, offer you the job!

It is always hard to answer a negative question such as 'What is your greatest weakness?' or 'Tell me about something that went wrong during your PhD. How did you deal with it?'. This is where preparation is indispensible so that you have something ready and are not caught out stuttering your way through the answer. In these situations, provide a real example of where you may have demonstrated a weakness but be sure to turn your answer into a positive reply by explaining how you overcame difficulties and solved the problem, or what you learned from the experience.

The employer is not so much interested in the detail of your negative experiences as how you turned them into positive outcomes. He/she is also interested to see how well you understand yourself. None of us is perfect – if we thought we were we may be unbearable to work with! See Chapter 3 for more about self-awareness.

Thinking of questions to ask the interviewer

People are often worried about the part of the interview (usually at the end) where you are invited to put forward your own questions. However, it is likely that questions will have come to mind during your interview. For example, you may want further clarification about the job role or the company. You can ask questions such as the following.

- What kinds of professional development opportunities are there?
- How do you see the company developing over the next 5 years in the light of new technologies such as xxx?
- Do you see this role/department expanding in the next few years?

It is not essential that you find questions to ask if you do not have any. Say: 'No, the interview has covered everything I needed to know'. Unless invited, don't be tempted to ask about holidays, salary and other personal elements of the job at this stage – negotiating these terms is only applicable if you get offered the job.

Preparation

When presenting a paper or poster at a conference, most people are more anxious about the prospect of the questions than the presentation itself. This is usually because the presentation and its content are under your control but the questioning is not. The idea of not being able to answer a question in front of a room full of people fills most of us with dread. Translate these circumstances to interview and the thought that our minds might go blank under the pressure of the situation is just as daunting.

In reality, the questioning is usually never as bad as you think it's going to be and, with some anticipation, you can prepare for the more obvious questions (Box 7.6). Equally, although you can never be certain exactly what questions you will be asked at interview, finding clues to some of them is quite simple and straightforward. For example, you can revise the advertisement and associated job description, review your application and the organisation/research group. As with an exam, you can

---

Box 7.6 Golden rules

- Always prepare in advance of the interview: research the company/research group, revise the job description and your application/CV.
- Try to predict the questions which will be asked, including the less obvious ones (e.g. they may ask you about your opinion on some policy point or a recent development associated with the industry).
- Practise your interview technique. This might include a mock interview with a professional adviser, colleague or friend. It could involve practising a presentation or maybe you need to try out some tests and exercises in preparation for an assessment centre.
- If you are travelling some distance to your interview, aim to arrive 30 minutes before it is due to start in case of unexpected delays during your journey.
- Dress appropriately according to the profession/company. The research environment tends to be informal so smart/casual wear is the norm for postdoctoral interviews. For more senior roles or management posts, smart dress is essential. As a rule, aim to be as, or more, smartly dressed than your interviewer.
- Take all relevant information with you, e.g. the job description, your application and covering letter and any other documents you think will be of interest to the interviewer.
- You are being assessed during every part of the process (including preinterview or postinterview activities, socials or meetings with other members of the department) so be nice to everyone!.
- Portray a friendly and confident disposition and, depending on cultural nuances, gauge greetings such as hand shaking and direct eye contact from the interviewer/ interview panel.
- Smile, look interested and don't gesticulate too much.
- Answer the questions clearly and succinctly. This is harder for broad questions such as 'Why do you want this job?', so practice is essential.
- Take a crib sheet with you to preview ahead of the interview, as you might do before an exam.
- Provide specific examples and evidence to back up each of your answers.
- If your main reason for applying for the post is personal or so that you can stay in a particular geographical area, don't use this as a primary reason when asked. You may have had to make a compromise but you will still have valid reasons for choosing this post over other types of jobs.
- Never ask about salary, holidays and other benefits during your interview unless specifically invited to. Wait and see if you are offered the job first!

usually predict many of the questions ahead of the interview and prepare your answers as far as possible. Thinking on your feet when questions are unexpected is more tricky but by this stage of the interview it is likely that you will have settled into its rhythm and your mind will be more agile. If not, it is best to be honest and admit you don't know the answer or to ask for the question to be clarified.

If you are offered the job

Congratulations – time to celebrate! However, being offered the job is not the end of the process. You may now need to negotiate the terms and conditions of the job before you sign your contract. This may be to agree a start date and where on the salary scale you will be appointed. Holidays, pension and health benefits may also be included in a company package which you need to agree (although many of these will be non-negotiable).

If you are taking on an academic research post, you may need to negotiate for office/lab space, equipment, teaching commitments, budgets, etc. Evidence suggests that men are better at this than women (Babcock & Laschever 2008). This process could take time and delicate bargaining so it is worth getting some advice from your supervisor or mentor before you sign your contract.

You need to be happy with your overall package – be prepared to back down on some things or agree that your position will be reviewed 6 months to a year into your contract when you have had time to prove yourself. Whatever aspect you are negotiating on, make sure you back up your reasoning with examples and facts which underpin your claim – this will add weight to your argument and objectify the discussion.

Sometimes an interview can show you that the job or company is not right for you. Therefore, if you are offered the job, you may decide to turn it down. Try to do this as soon as possible so that the employer is able to offer the job to the second candidate or readvertise the position. Rejecting a job offer can sometimes be a relief and demonstrate to you what you really want to do – a case of the negative revealing the positive.

If you are not offered the job

Any rejection in life is hard to take so don't be ashamed or embarrassed about your feelings of disappointment – it's only natural. There will be a number of reasons for you not being selected for the job.

- You gave the impression at interview that you were not interested in the job.
- Your interview technique was poor.
- You were poorly prepared or overly nervous.
- There was someone already lined up for the job.
- You were not the best candidate.

Whatever the reason, this is the time to draw upon your career planning 'tools' (see Chapter 2) and try to stay positive, motivated and persistent. Reflect on and analyse the interview and write down some notes to recall areas which you could improve upon. Perhaps there were gaps in your experience that need to be addressed, maybe your research into the company was lacking, or you could have answered the questions better. Your interview performance may have been near-perfect but you were unlucky to be beaten into second place by another candidate who was equally deserving. You can ask for feedback to discover why you weren't successful but you may find companies give you a generic answer such as 'The field was very strong'.

The important lesson is to move on and look to the future and your next opportunity. Ironically, it is sometimes the job people think is perfect for them which they miss out on. Perhaps because they want it too much, they try too hard. If this happens to you, with any luck you have other options brewing and you are researching and applying for alternative positions. If you think your interview technique needs improvement, consider doing a practice interview with a careers adviser or consultant at your university, or privately. A colleague or friend may agree to role play with you and give you constructive feedback. Whatever you do, don't despair and don't give up. Speak to people, get help, review your strategy and move on. You have proved you can reach the interview stage of the process which means your applications are effective; you just need to hone your interview technique and you'll be there!

Conclusion

This chapter provides advice on the essentials of successful interview technique with a guide to the kinds of format and questions you are likely to encounter. You could find yourself in an interview for all kinds of reasons, not just for a job application. It may be for a voluntary post or internship. It could be a casual visit to a company to find out about possible openings or work shadowing opportunities. Whatever the reason or circumstances, the same rules apply, even though the situation may not feel so serious.

Always prepare and dress appropriately to show your intentions are serious and try to anticipate the likely line of questioning and respond with confidence and clarity. There are many books and other media offering more in depth advice and information on interview technique (see Appendix 4). University careers services may organise workshops or offer individual practice interviews.

Reference

Babcock L, Laschever S. (2008) *Women Don't Ask. Negotiation and the gender divide*. Bantam: New York.

CHAPTER 8

Decision making and action planning

'Actions speak louder than words', so the saying goes. Thinking things through and talking about them is one thing. Turning your thoughts into reality by making a decision and taking action is another and involves making a commitment.

In this chapter I have joined together 'decision making' and 'action planning' as they are inextricably linked in the career planning process (see the DOTS model in Chapter 2). The result of decision making is the need for action. How soon you implement your decision depends on factors such as the urgency of the situation, the number of options you have, your motivations and personality (see Chapter 3). Sometimes it is only when we feel discomfort that we start to take action. For example, being in the 'wrong' job can cause stress and anxiety. Equally, you can be in the 'right' job but are coming to the end of your contract, which may create similar negative feelings. Conversely, if you are starting a new job or making any type of career transition following a proactive and positive decision, the change will be more satisfying and may give you an initial surge of 'elation' (Box 8.1).

Careers in research

At this stage in your career you may be weighing up which career path(s) to choose. Should you stay in academia or consider other options? Maybe you are thinking about moving to a career completely unrelated to your discipline. For many doctoral students and postdoctoral researchers, a research career is what you most desire. You are passionate about your research topic, you like the research environment and would prefer to remain in it for the duration of your career; it may be what you committed to when you made the decision to do a PhD.

If you are considering aiming for one of the 'top' academic jobs in a university, bear in mind that, in most developed countries, only a small proportion of researchers will obtain a permanent senior academic position (see Chapter 1, page 1). As with most professions, the career progression profile for academia is pyramid shaped, with many more people at the lower levels and a tiny proportion at the top. This profile has been described by Dr Jennifer Rohn of University College London, in Rumsby (2011), as being more like a 'spike on a flat plain', with a burgeoning postdoctoral community chasing very few permanent academic positions.

Box 8.1   Transition and change

Change can be an exciting but also a daunting and difficult process. As defined by Adams et al. (1976), change (or transition) is a 'discontinuity' in a person's life space. They formulated a seven-stage model of the change process (see Fig. 8.1) which charts the effects of transition on a person's self-esteem and resultant behaviour. As stated by Blair (2000), 'A discontinuity requires alteration to routine, habit and the taken-for-granted configuration of occupations. It requires personal awareness and recognition of the event and new responses to deal with the results of the discontinuity. This is frequently painful and can result in protracted denial'.

Adams et al. (1976) identified four types of change: predictable voluntary, predictable involuntary, unpredictable voluntary and unpredictable involuntary, all of which come with differing degrees of stress to the individual. Therefore, if you are feeling depressed or immobilised by the prospect of moving on after the end of your current contract, bear in mind that this is a normal reaction to change, especially if it is involuntary. You can overcome this negativity, to some extent, by taking action such as addressing your personal and professional development needs to improve your employability (see Chapter 5). It need only be something small – attending a staff development course, making an appointment to see a careers adviser, discussing your situation with your supervisor or mentor, or starting to peruse the job market to see if anything catches your attention.

Moving to a different job, whatever the circumstances, usually requires getting to know new people, adapting to a new work culture and even moving to another town or country. If this is your first move in many years, it may accentuate your reaction to change. Bearing in mind the transition model in Figure 8.1 will help you understand the process and your response to it.

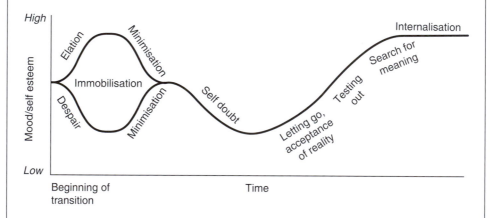

Fig. 8.1 Mood/self-esteem changes during change/transition showing an initial phase of elation or depression, depending on a person's motivation to change, followed by a period of self-doubt. As acceptance starts to prevail, it is possible to start letting go of the past and to move forward. These stages can last for any length of time and recur. The model is relevant to any change, including careers but also other life events. Reproduced from Adams et al. (1976).

Other career options

For those thinking of changing careers paths into a non-research job, you may not currently possess the experience, qualifications and skills required for entry into other professions at a level you might expect. Consider the situation for someone wanting a mid-career change from accountancy into academic research. They would need to retrain to gain experience and a PhD in order to enter the profession and progress. They may have to take an initial step back in terms of their position and salary. Similarly, for those considering moving from research into alternative careers, your career trajectory may 'wobble' initially as you start to adapt and shape it differently. Your training is research specific so moving onto a different career track may not be straightforward. Time spent conducting research into other career sectors and undertaking extra activities, such as those specified in Chapter 5, to enhance your employability may be more beneficial than publishing your work. If you prepare and position yourself favourably by networking, marketing your highly transferable skills and taking control of your career, you should be able to smooth your transition. Many people have transferred very successfully into a wide range of alternative careers (see career narratives in Appendix 1). Sometimes the thought of doing something is worse than the reality!

Decision making

In order to take action and move your career forward, you will need to make decisions. You have already been making lots of decisions which have brought you to where you are now. Consider the circumstances of some of these decisions. Was it a proactive or a passive process? Was it part of a structured career planning strategy or was it a 'happenstance' moment where you took advantage of a chance opportunity which came your way?

Decisions infiltrate the entire career planning process. Even what appear to be quite minor decisions during the course of your PhD or postdoctoral research can exert an influence on your career. The career narratives in Appendix 1 illustrate such cases with people citing chance events in their lives that have proved to be a turning point in their career. Even attending a departmental seminar or emailing someone to find out more about their work can contribute to your career development or progression. Meanwhile, the more significant decisions still loom large. Which career areas and jobs are suited to you? Should you continue to pursue a career in academic research? When should you start your job search? Which offers should you accept? Getting used to being proactive about making decisions will provide practice for these more serious life decisions: if you look after the small decisions, the larger ones should take care of themselves.

Decision-making styles

Decision making is one of the cornerstones of the DOTS career planning model (see Chapter 2). As a careers adviser, I assist people, through structured coaching and counselling, to clarify their ideas and thoughts and to help them to make effective and informed decisions. We all make decisions in different ways according to our own personalities and preferences. Some of the main styles are listed in Table 8.1.

Generally, people are encouraged to use the rational method in decision making. That is, to examine the facts, then weigh up the advantages and disadvantages before settling on a plan of action. On the other hand, decision-making styles such as

Table 8.1 Decision-making styles.

Decision style	Description
Rational	Logical, informed and dispassionate. Weigh up all the relevant factors before making a decision
Intuitive	Go by your instinct or your 'gut feeling'
Impulsive	Take decisions spontaneously, driven by initial and surface feelings
Procrastination	Tend to think about things too much and put off making decisions
Compliant	A passive approach, allowing others to make decisions for you
Overanxious	Become overly worried about the consequences of your decision and inertia can set in
Fatalistic	Fate will decide – tends to be a passive attitude

procrastination, compliant, overanxious and fatalistic are the least effective as they tend to be associated with a combination of passive, negative and inert behaviours which do not lend themselves to a proactive and positive approach. Intuitive and impulsive approaches to decision making, although seemingly not immediately functional, can be valuable in situations where an unexpected event has arisen and swift action is needed. They may fit better with a 'planned happenstance' approach to career planning (see Chapter 2), as intuitive and impulsive decisions tend to involve risk and are driven by curiosity and optimism.

Recognising and acting on chance opportunities when they arise can be the difference between getting on track with your career and missing out on a golden opportunity. Perhaps you can think of a time when an intuitive or impulsive decision you have made has led to a positive outcome. The rational decision-making approach is more ordered and gives you a framework on which to plan your career so that when unplanned events come up, you are ready to act. Combining rational thinking with intuitive or impulsive decision making can provide a highly effective attitude which will serve you well in a dynamic and unpredictable world.

Taking action

Nowadays, organisations need to be flexible and able to respond to new developments if they are to operate effectively in our fast-paced global economy. Tied into this, people in the workforce need to be self-reliant and adaptable to maintain their employability (see Chapter 5). In response to this changing world of work, Hall (2004) identified a new type of career, which he described as 'protean'. In contrast to the traditional work culture, where it was the organisation which was in charge of career development and progression, the protean career is one where the individual is in control. This fact is very pertinent in the 21st century, where companies are increasingly outsourcing and expecting their workforce to readily adapt to changes. Academics are required, more than most, to be autonomous and take control of their careers by seeking their own funding in order to sustain their research portfolio (and career). Hall cites 'proactivity' as being the central factor to enable a person to be more protean. He proposes two personal attributes, or metacompetencies, as

Table 8.2 A combination of high self-awareness and high adaptability provides fundamental personal 'metacompetencies' for taking a proactive approach to career planning (adapted from Hall 1984).

Effectiveness	High adaptability	Low adaptability
High self-awareness	**Proactivity**: 'smart' performance	**Paralysis**: blocking; avoidance
Low self-awareness	**Reactivity**: chameleon behaviour	**Rigidity**: performing to order

being vital to achieve proactive behaviour: (1) well-developed self-awareness, and (2) a high level of adaptability. Without this combination, resultant behaviour can range from passive, compliant (low self-awareness, high adaptability) to rigidity (low self-awareness, low adaptability; see Table 8.2). Translated into the decision-making process, this implies that those who are more aware and knowledgeable about themselves (see Chapter 3) and who are willing to adapt to change are more likely to make informed decisions about their career and maintain their employability.

A reactive and compliant approach may be effective in the short term, resulting in research contract extensions but it can, ultimately, lead your career to a dead end. This applies to seemingly small decisions and actions which you take on a daily basis. For example, if your department has organised a careers workshop, seminar or internal conference, consider seriously whether you should attend. It could benefit you and your career, even if only in a small way. It is not necessarily the most talented and most highly qualified people who succeed in their field, it is those who take a proactive attitude who are more likely to reap the greatest rewards.

Throughout this book I have offered ideas and suggestions for taking action with regard to career planning: improving your self-awareness (Chapter 3); analysing the job market and networking (Chapter 4); enhancing your employability (Chapter 5); and marketing yourself to employers (Chapters 6 and 7). But *what* action will you take and *when* will you take it? At the end of my career development workshops, I often ask participants to write down three actions they intend to take as a result of what they have learned. I ask them to place the postcard in a self-addressed envelope which I then collect in. Three months later I post the letters out to everyone to remind them of their intentions. I have attended workshops where I have also had to do this task and, from my own experience, reviewing my intended actions 3 months on can be very motivating.

Turning decisions into action

Personal development planners such as the Researcher Development Framework (Vitae 2011) and the NPA Core Competencies, which are tailored specifically for postdoctoral researchers and doctoral students, enable you to keep a record of your skills and experiences so that you can monitor your personal and professional development (see Chapter 3). Built into these personal planning tools are opportunities for you to identify development needs and state the actions you will take to achieve them. There are also resources for you to follow up to help you find out more about developing individual competencies.

Coaching models are available to help you to clarify and structure your thoughts in order to make decisions and turn them into actions. For example, the GROW model (Whitmore 2002) defines the fundamental stages leading to a decision and action as:

- **Goal** – what do you want to achieve (in the short or long term)?
- **Reality** – what is your current situation? Are there obstacles to be overcome?
- **Options** – what options do you have in order to move things forward?
- **Will** – what will you do? What action will you take?

Questions to consider within the context of the GROW model include the following.

- *What is your GOAL? What do you want to achieve?* Are you planning for an academic career? Are you planning to do research outside academia? Are you considering a science-related career but not research? Do you have a contingency plan in case your main plan does not work out?
- *What is the REALITY of your situation at the moment?* Do you have enough papers to realise an academic career? Are you enjoying your research environment but not the research itself? Do you know which skills and qualifications you need to enter an alternative career which interests you? Do you have contacts and a network to help you? Do you need to stay in this town/part of the country due to family commitments?
- *What are your OPTIONS in the light of your situation?* Maybe you need to write more papers or speak to your supervisor about taking on more responsibilities which will improve your academic career prospects. If you want to remain in a university environment, you could look at your university vacancy list or examine other job sites which list academic-related posts. You could peruse the wider job market by looking at other job sites. If you have a career in mind, you may consider taking a course or doing an internship to improve your competitiveness. Perhaps joining a social network and improving your visibility will help you to find out about options or put you into contact with people who can help you.
- *What WILL you do? You have lots of options but what action will you take?* It is more productive to focus on a few options and put them into action than to be overwhelmed by many, which may lead to inertia and procrastination. Consider all your options and then focus on those which are achievable or most important in terms of moving your career forward.

Writing a plan of action

Tables 8.3 and 8.4 give example structures for writing an action plan. You may have a more personalised version you prefer to use or you can make use of recognised tools.

Talking to a professional careers adviser or coach can help to determine your priorities, enabling you to make more effective and informed decisions. Alternatively, writing down your thoughts and ideas or talking them over with a friend, colleague or mentor is also a helpful way to clarify your options. The skills audit, personality and value assessments in Chapter 3 may help you to focus on the areas of your research you enjoy, giving you clues to where your career interests may lie.

Table 8.3 Personal development plan, structure 1 (with examples). Based on the GROW model (Whitmore 2002).

Stage	Example 1	Example 2	Example 3
Goal	To get my results published in xxx journal this year	To improve my research profile	Get better connected within my research field
Reality	Although I have enough results to publish, they need collating and analysing. I'm not confident writing the paper on my own	I'm not well known and have only presented posters at conferences	I feel quite isolated from my research field as there are only three of us working in my area in the university
Options	Talk to my supervisor and discuss a plan of action. Agree which journal to target and look at the guidelines	Look for conferences to target. Start to network and promote my research more widely	Look at ways to improve my research profile by getting more involved in internal and external events and activities
Will	Attend an internal 'getting published' course to help improve my skills. Make a start on the introduction and analysing the results. Then discuss with my supervisor before writing the rest of the paper	Talk to supervisor about submitting an abstract to a conference. Sign up to a social networking course. Talk to others who are already blogging	Join one or two learned societies. Attend seminars in the department. Join the postdoc association and journals club

Table 8.4 Personal development plan, structure 2 (with examples).

What development do I need to do?	Where will I do it?	When will I do it?	How will I do it?	What will the outcomes be?
Social networking	Computer	Now	Join Twitter and LinkedIn. Talk to xx who has a blog	Increase my visibility and make more contacts. Hopefully access the 'hidden' job market
Face-to-face networking	Conference	July	Identify people ahead of the meeting and aim to talk to them	Make it easier to approach people. Come away with useful contacts. Maybe set up a lab visit
Increase my awareness of the job market	Computer, department, conferences	Now	Look at job sites. Networking	Widen my knowledge of other career options and maybe even find something unexpectedly
Write my CV	Computer	Now	Start brainstorming my career history and structuring the information	Will be ready to make applications if they come up

Conclusion

Effective decision making and action planning will help you to master your career so that you avoid compliance. Recognised theoretical models, e.g. the protean career (Hall 2004), explain the attitudes needed to thrive in the modern (unpredictable) work culture and show you how to take a proactive approach. As researchers, many of you may have moved seamlessly from your college degree to a PhD and on to a postdoctoral post with little experience of formal application procedures or a panel interviews. You may not have had work experience beyond university and so the prospect of a different working environment is a little daunting.

You have the power to take control of your career. Making decisions and taking action (however minor) will help you to stay motivated and enhance your employability. The 'big' decisions in your life will be easier to make if you have already injected some momentum into your career planning.

References

Adams J, Hayes J, Hopson B. (1976) *Transition – understanding and managing personal change*. John Wiley & Sons, Inc. Originally published by Martin Robertson, London.

Blair SEE. (2000) The centrality of occupation during life transitions. *British Journal of Occupational Therapy* 63(5), 231–7.

Hall DT. (2004) The protean career: a quarter-century journey. *Journal of Vocational Behavior* 65, p. 1–13. ©2003 Elsevier Inc.

Law B, Watts AG. (1977) *Schools, Careers and Community*. London: Church Information Office.

Mitchell KE, Lewin AS, Krumboltz JB. (1999) Planned happenstance. Constructing unexpected career opportunities. *Journal of Counselling and Development* 17(2), 115–24.

NPA Core Competencies. Available from: www.nationalpostdoc.org/competencies.

Rumsby M. (2011) *Research Careers – opening the discussion*. London: Royal Society. Available from: http://bit.ly/AyyAHZ.

Vitae. (2011) Researcher Development Framework. Available from: www.vitae.ac.uk/RDF

Whitmore J. (2002) *Coaching for Performance*, 3rd edn. London: Nicholas Brealey Publishing.

Afterword

Chapter 8 concludes this book and brings together all the accumulated advice and information from the other chapters and appendices.

- Chapter 1 introduces the book and cites key policies which have shaped and influenced the evolution of career development support for research bioscientists. It establishes the aims of the book and gives a brief description of each of the chapters.
- In Chapter 2, two career planning models – the DOTS model and the 'planned happenstance' theory – are presented, which underpin much of the content of this book. They illustrate how structure and flexibility complement each other to allow you to react to planned and unplanned events. Many of the career narratives in Appendix 1 demonstrate how chance, as well as good planning, has played a part in people's careers.
- Chapter 3 introduces you to the importance of self-awareness and how knowing yourself can help you to assess your career options more accurately. Most of us define ourselves in terms of our knowledge and interests. This chapter focuses on three other elements of 'self' – skills, personality and values – and provides exercises for you to complete to enhance your self-awareness.
- Knowledge of the job market and how to access unadvertised positions is a vital component of career planning, which is described at length in Chapter 4. In the first part, examples of job vacancies illustrate the similarities and differences between job specifications in a variety of industries. The second part of this chapter highlights the importance of networking as a route into the 'hidden' job market and suggests ways for you to increase your personal visibility, including using social media (see Appendix 2).
- Being self-reliant and taking control of your career development is inextricably linked to the concept of the 'protean' career described in Chapter 8. In Chapter 5,

Career Planning for Research Bioscientists, First Edition. Sarah Blackford.
© 2013 Sarah Blackford. Published 2013 by Blackwell Publishing Ltd.

examples of personal and professional development activities are provided to help you to enhance your employability.

- Transition from your current post to your next job, whether it is within the same career area or to a different occupational field, usually involves an application procedure. Chapters 6 and 7 show you how to write an effective application and perform well at interview. Clear, structured and relevant communication is at the heart of this process and the guidelines and advice in these chapters aim to help you be successful and stand out from the crowd in a highly competitive job market. Example CVs in Appendix 3 show you how the content of your CV can be structured according to different job specifications and a sample covering letter provides a format for introducing your CV.

- Chapter 8 shows you how to make effective decisions so that you will be confident in the actions you take to achieve a successful and fulfilling career.

All the chapters in this book play a part in helping you to plan your career. Take advantage of the advice and information in them, use the additional resources in Appendix 4 (stay up to date with my blog, www.biosciencecareers.org) and remember to stay positive.

The future starts here!

Career narratives

Twenty career narratives are presented from PhD-qualified bioscientists working in a range of careers. Each narrative is split into three parts: career facts, career factors and a commentary. 'Career facts' describes each of the job roles and gives an account of the person's background, charting their transition from PhD student to their current job. The 'Career factors' section cites significant aspects associated with career planning such as key decisions, events and personal action people took in order to arrive in their current job. Many of these factors have been described in the book, such as the importance of personal development, networking, being proactive and taking a strategic approach to career planning. I provide an objective commentary at the end of each career narrative, picking out noteworthy points of interest and offering general careers advice and guidance. Resources related to these narratives, such as websites advertising jobs in particular sectors, sources of funding and mentoring, are detailed in Appendix 4.

Summarised list of career narratives

Academia

1 Professor (US university)
2 Research fellow (Austrian university)
3 Lecturer (UK university)

Industry

4 Lead scientist (large agribiotech company)
5 Scientific team leader (large pharma company)
6 Product and R&D manager (small biotech company)
7 Scientific adviser (protein services company)
8 Company director (small immunoreagent company)

Science communication

9 Senior medical writer (medical communications agency)
10 Science editor (freelance)
11 Features editor (online education, scientific journal)

Career Planning for Research Bioscientists, First Edition. Sarah Blackford.
© 2013 Sarah Blackford. Published 2013 by Blackwell Publishing Ltd.

12 Teacher training co-ordinator (public research institute)
13 Science journalist (freelance)
14 Policy officer (governmental/charitable organisations)

Specialist and technical administration

15 Clinical trials co-ordinator (university cancer centre)
16 Science administrator (European funding organisation)
17 Patent examiner (European patent office)

Non-science

18 Healthcare analyst (self-employed partnership)
19 Technology consultant (international consultancy company)
20 Sound engineer (production services provider)

1. *Michel*: professor, US university

Career facts

Job description

In my role as full professor, I have been privileged to be involved in cutting-edge research, to have seen many of my research ideas tested and to have worked with many interesting and highly talented individuals. All of this has taken place within a stimulating, autonomous and open-minded working environment with the flexibility and freedom to be creative, to inspire (and be inspired by) young people, and to be challenged intellectually. After 25 years of running an active, well-funded research lab, I am seeking new challenges in order to let the next generation of researchers come through. I can look back now and relate my experiences, which have brought me to this point in a career that I have thoroughly enjoyed and still love.

Background

I did my PhD in cellular and molecular biology in Michigan after which I went to Seattle to undertake a postdoctoral position for 4.5 years. At the age of 32, I then secured an assistant professorship at my current university in Minnesota. This is an unusually young age to get a faculty position but that was in the late 1980s. Nowadays, late 30s is a more normal age; most researchers do one long or two postdoctoral training fellowships before they are hired in as an assistant professor. Assistant professors are not tenured, and it is during the next several years that they establish their credentials for tenure by securing significant research funding and publishing scholarly research articles. Then, at the end of the fourth/ beginning of the fifth year, they need to submit a dossier for review by several committees, the dean of their college, the president of the university, and ultimately the Board of Regents for the university. This review takes about a year. If their scholarship is deemed satisfactory, they get promoted to associate professor with tenure. If the scholarship is considered unsatisfactory, then the annual contract is terminated after 1 year, and the unsuccessful faculty member will have to leave the university and look for a post at another university or leave academia altogether. Even if one does secure the position, however, it takes between 6 and 10 years (or longer) to reach full professor. However, the success rate is quite high (95%) at most universities as those unlikely to make it often opt not to go through the promotion process.

A fundamental difference between being employed as an assistant professor and associate professor is, apart from other aspects, the teaching and mentoring load. As an assistant professor, you are given practically no teaching duties during the first couple of years, leaving you free to pursue your research with vigour. This increases during the next 2 years as teaching credentials are one of the criteria for promotion. In contrast, an associate professor may be given up to 25 hours teaching per year, which is a heavy commitment considering they are also trying to develop their research career towards a full professorship. (I should add that this load is typical for faculty in a medical school; faculty in other colleges usually has a much heavier load).

For career advancement to full professor, one will also need to be considering important career factors such as gaining an international reputation, getting onto journal refereeing or editorial boards, as well as departmental commitments such as mentoring students, committee work and administration. For my part, I was on the promotions and tenure committee for the medical school, serving as the chair for 5 years. I was also on the faculty senate and medical school curriculum boards, as well as national committees, e.g. grant review committees, so-called study sections. Administration and committee work are less appealing to me personally than research-related roles, but I learned a lot from my involvement in these services. As all faculty are required to provide 'service' to the university in addition to research and teaching, it is important to select committee work that is either of interest to you and/or that is of value to you. For example, I chose to serve extensively on grant review committees both locally and nationally because I found it interesting to read about the latest ideas and techniques, because it helped me keep my research programme at the forefront of the field, and because it has a measurable and important outcome. Learn to say yes only if you think the activity will be useful to you or is of personal interest, otherwise you will be quickly swamped with too much work (and no motivation to do much of it).

Career factors

You have to be strategic if you are planning on an academic career. You need to choose an area of research which interests you but is also receiving plenty of government funding. In this way, you will be publishing regularly and will have the opportunity to move around easily to work in other labs – mobility is very important at the early stages of your career. Moreover, other aspects are important such as setting up your laboratory. For my first tenured post, I moved into a pre-existing lab with plenty of equipment and so only had to order a small amount of additional items to start getting my experiments up and running. However, if you are faced with a completely empty room with an equipment fund of only $800,000 you are going to have to start applying for grants in the first couple of years, especially if you want personnel such as a technician. You can make use of graduate students to help with your research, but you will need to start eyeing up the 'big money' if you are going to make progress in your research. You can gain experience with writing grants while you are still a postdoctoral fellow by writing for fellowships – developing this skill early on in your career is vital. Even once you have started to employ students/researchers, you are still the most trained person in the lab so you may need to spend time at the bench for much of the first 5 years. Another consideration is the selection of a mentor to provide guidance to you. Too often, it is tempting to go it alone, thinking you know what is needed; however, the advice of someone intimately familiar with the politics and the writing of grants is invaluable.

In the beginning, you will likely employ the services of a doctoral student and one to two undergraduate students (e.g. work study students). This is realistically the most you can supervise and still do research yourself. Then, by your third year you will start to bring in a postdoctoral researcher so you will have four people in the group. After this, it is advisable to hire slowly – one person per year – so that you can reasonably manage the group's growth. Different labs have different management styles according to those who are leading them. I had never managed anyone when I started up my own lab – I was good at having ideas, doing experiments and making presentations. Even hiring good people is difficult when you have no experience. Again, a mentor can be valuable in making these decisions and selections.

Once you are established you can relax (a little) and concentrate on doing more peripheral activities whilst managing your lab, such as focusing on teaching and mentoring activities. At a university, teaching is a primary mission and your reputation is also dependent on how well you teach. Of course, if you want to make policy decisions and guide the growth of a department or college or do editorial or national committee service, then you can also divert time into those activities after reaching the associate and/or full professor levels.

Commentary

Faculty positions work differently in different countries. The terminology and priorities associated with them can vary significantly, so it is important that you are aware of this when you are making applications. A lectureship and professorship in the USA and in the UK, for example, come with different sets of responsibilities and even expectations for tenure (see Chapter 4, page 40, for more information on academic posts and career tracks). Whatever country you are based in, you will need to be highly strategic and focused in order to achieve a full academic career. Although, as Michel points out, it can be a highly challenging and enjoyable career, it is also intensely competitive and demanding, especially at the earlier stages. Even when you do secure a faculty position, you are still under pressure to continually produce results for publication, apply for funding and juggle administrative, teaching and any number of other outside activities of benefit to your career advancement. If you are aiming for an academic career and you think, at any stage along the way, that you may not achieve it, you must seriously consider a contingency plan – particularly when you are at the assistant professor (second/third postdoctoral) stage. Otherwise you may find yourself in a series of temporary contracts. Academic careers can be highly rewarding and stimulating but bear in mind that the chances of getting to the 'top' these days can be slim and require dedication (see Chapter 1, page 1). On the plus side, those who make it love what they do and wouldn't have traded a life of exploration and learning for any other career.

Career facts

Job description

I started my Women in Science Support Fellowship in March 2008 and it is due to run until 2014. This 6-year fellowship is aimed at women from the start of their career up to 'habilitation' (see Chapter 4, Box 4.3) and can last between 3 and 6 years, providing the opportunity to undertake professorial duties and, of course, putting you in a better position to secure a faculty position. I've already been involved in departmental activities such as student supervision, teaching and organising seminars which is enriching my research experience and, hopefully, enhancing my career prospects. I have just returned from a 6-month research visit to a university lab in the UK, having been approached by a researcher following a talk I gave at a conference. This has further strengthened my position and we are now collaborating on a new project.

Background

I was already applying for my current fellowship a year before my previous (second) postdoctoral position was due to finish, having used the same tactic for securing my first postdoc too. In fact, my PhD overlapped with this first postdoctoral position and was a result of a funding application I made with my supervisor to the Austrian Science Programme. So in 2005 I found myself working on two projects at the same time which helped to enhance my research output as well as my personal project and time management skills! I continued working on several overlapping projects in Austria and abroad and this has recently earned me a fellowship from the Chinese Academy of Sciences.

Career factors

Be proactive

I seem to have spent my career in 3-year blocks so far, always looking ahead and applying for funding to set me on my way again. I'm not a person to sit around waiting for someone else to give me a hand and have always taken a proactive approach to my life. Our university runs funding information workshops which have helped me keep abreast of the pots of money available and my supervisor's mentoring has helped me to write them. At the time when I was about to finish my PhD I had the opportunity to take part in very useful career workshops which proved useful to me later on in my research career.

Don't be afraid to put yourself forward to present a talk – it was a daunting prospect to stand up and present my first conference talk when I was a PhD student; little did I imagine that one of the reviewers of my first postdoctoral fellowship application had seen this talk which influenced his opinion of my abilities.

One of my most fruitful activities has been volunteering to organise the departmental seminar series. Although there is no credit or obvious benefit attached to this 'job', it has put me in touch with many eminent international researchers whom I have since met at conferences and generally extended my network of contacts.

Get training

During my first postdoc I attended a 1-week teacher training workshop which gave me the confidence and skills to offer to lecture in the department (unpaid but great experience). If you get the chance to attend summer courses, these are valuable ways to learn new skills and do some more networking with research scientists in (usually) other disciplines. I have attended two in the US: one on the technical aspects of ageing (a week in California) and another on molecular methodology (a month in Woods Hole). For most courses, there is funding available for travel and living costs but be prepared to self-fund if you think the course will give you an advantage. For me, attendance at both courses was financed with fellowships from US funding organisations. Be alert and watch out for flyers and announcements for courses and once you've identified a suitable training event, be proactive and ask the organisers for funding possibilities. In my case, my home university offers some support if a proposal convinces them that attendance at the course will be of importance. Also, other organisations and learned societies offer financial support for lab visits and conference attendance.

Commentary

Teresa has been very proactive during her career and has always kept an eye on the future. In this way she's been able to prepare herself for her inevitable next career transition. Keeping a watchful eye on the funding landscape and relevant upcoming conferences is one thing, but taking action is crucial so that you do not miss out on opportunities to achieve research independence and make valuable contacts.

3. *John*: lecturer, UK university

Career facts

Job description

My post is a tenured academic lectureship which comprises three primary activities: research, teaching and administration. About 40% of my time is spent on research, mainly writing grants and papers. I have six grants at the moment with two postdoctoral researchers and three PhD students working on the projects. Two Master's students usually join us each year for 6 months. My role is to monitor the results being generated by the group and to come up with strategies and ideas to take the research forward. I am collaborating with a number of research groups around the world and we consult on a regular basis to discuss current work and generate new proposals. My collaborators are widespread and include groups based in Malaysia, France and USA. I keep in touch with them via email and manage occasional visits. I also keep an eye out for potential new partners to link with and submit collaborative proposals depending on the grants coming up. I do this by writing to groups based on papers I have seen, then negotiating and discussing ideas to see whether it would be useful to work together. Sometimes things work out well (and you can even become personal friends), other times they don't and you move on your separate ways.

My working week is long and includes evenings and weekends. This is probably true of most academics. There is a lot to do what with all the research and teaching responsibilities. I teach around 30 lectures per year with 15–20 practicals on top, which represents about 30% of my time. With this comes preparation and administration, exam setting, assessment, etc. My subject coverage is close enough to my own specialism which I enjoy, but far away enough that I am learning new things myself. I give postdoctorals the opportunity to teach if they want to, but I would say that activities such as presenting lectures at conferences, PhD supervision, reviewing papers and giving feedback provide enough evidence of potential teaching ability, as well as a broad knowledge of your field, and the ability to write good English. It is still funding and publications which dominate if you are to stand any chance of securing a tenured position, although teaching fellowships and short-term contract teaching lectureships do exist if you want to go down this career track.

The rest of my time is spent on administration within my department and on external activities. I am a journal editor which means I review papers, oversee the editorial aspects, make decisions and liaise with the other editors and publishers. I am a member of a few learned societies, for one of which I sit on council. This involves advising the president and organising the main meeting. I have also recently been appointed chair of their education and outreach committee. Another important activity is contributing to and attending conferences – remote communication is fine but face-to-face networking and socialising can build relationships faster and more effectively.

Networking is vital as research is so scattered these days, with groups working in your specialist area spread out all over the world. You may only find out about some grants through word of mouth so you need to be in touch with people in your field. Furthermore, specialised discussion lists and websites of relevant funding bodies and other organisations will also advertise opportunities so you need to be 'in the know' about all of these. Part of my work, I feel, is to mentor my postdoctorals to develop some of these associated skills to prepare them for academia, should they wish to pursue this career route. It is not beneficial for them to simply produce results and papers without the experience of making connections, building up their own network and moving towards independence. They may not be able to write the grant, or may need help from me, but being aware of the research funding landscape helps them to develop – and it helps me too! It would be hard for me to do this job without delegating as there's far too much work to do as an academic. I wish I had taken control of my own career and paid attention to developing my ideas and direction earlier. I got my academic career in the end, but the journey certainly wasn't easy or straightforward.

Background

Following my PhD and first 3-year postdoc, I took a year off to pursue personal interests and to stay in the same location. My first child had just been born so I was able to spend a lot of quality time with him. Following on from this, I managed to secure another 3-year postdoctoral research post, but by the end of it I was determined not to do another postdoc and to focus only on fellowships and permanent academic positions. I thought if I did another postdoc, I would get stuck in a rut of moving from one short-term contract to another and wanted to avoid this uncertain future. This led to me being unemployed for a year while I made application after application – over 20 in fact. I wrote grants for fellowships and as a named researcher on other people's grants. In retrospect, I should have thought about doing this 2 years into my previous postdoc, but I didn't think about it until the end of the contract. When I look back at my early applications for academic posts and fellowships, they were very poorly written and I can see why I didn't get called for interview. The rejection letters helped me to hone my skills! As the year progressed, my applications improved and I finally secured a fellowship and by then valuable lessons had been learned. I used my fellowship to good effect, meeting people and building up collaborations, finding a research direction and generating enough papers to apply more realistically for an academic post. Within 2 years I had applied, and been accepted for my current post.

Career factors

Be assertive

As a postdoctoral researcher, you may be pressured by your supervisor to take on areas of work which are of no interest to you or which may impede your progress. Of course, you want to show willing and help out if you can, but be discerning and only say yes to things which you think will benefit your career. If you are considering a non-academic career, associated responsibilities such as teaching and outreach may be beneficial to you. However, if you are aiming for an academic career, you need to think carefully; your contract is very short and you have limited time to do research and prove yourself. Supervising PhD students is fine and gives you managerial responsibilities but only if their work is connected in some way to yours; don't get sent off in research directions which are not of interest to you. The most effective way to assert yourself is to be sure of what you want. Write a plan and present it to your supervisor. Academia can be quite hierarchical but he/she should be impressed with your independent thinking and value your wish to take control of your research.

Sell yourself

They say you need commercial skills if you want to work in industry, but academic research demands that you routinely sell yourself and your science. Writing grants is like writing a business plan – you have to research the market, write your proposal and justify its 'profitability' and your ability through published papers and knowledge transfer (for example to industry, the public and other stakeholders). Equally, you have to know when to give up on a proposal. Don't get too wedded to your research ideas – they may be brilliant but if no one will fund them, move on. I have grants for projects I love and others which aren't as appealing, but you have to learn to make compromises and follow the money to have a sustainable academic career.

Commentary

John has provided really invaluable advice here. Primarily, he stresses the need to gain research independence as soon as possible if you are a would-be academic. Also, anticipate and prepare for the end of each contract by starting to build your own network and familiarising yourself with the research field and sources of funding. Don't be afraid to approach your supervisor and voice your ideas in relation to the research project and its direction. Most academics are happy for their postdoc to take control as long as they are being productive.

4. *Andrew*: lead scientist, protein design team, large agribiotechnology company

Career facts

Job description

In my current role I lead a small team of three, designing and improving protein function. The work is highly interdisciplinary and involves interacting with many other groups with specialist expertise in areas such as biochemistry and entomology, so there is a lot of communication and co-operation required. When I started in the company, my work was more aligned with that of a postdoc. Now in my fourth year, I have a more senior managerial position which is more hands-off in terms of practical work but involves a lot of planning, leading and steering projects. I spend a lot of time preparing for meetings, reporting on progress and communicating with other expert groups. I am learning more and more about the business end of things and I also contribute to the company's internal conferences.

Background

Following a postdoc position which lasted for 2.5 years, I decided not to pursue my career any deeper into academia. Having witnessed the long hours and weekend working practices of an assistant professor who was trying to get tenure, I wanted a career with a more even work–life balance. This was the main driver for my move into industry – more stable hours, a more structured career path (and a little more money). I applied for my current job and was offered an interview in which it was clear that my technical skills and specialist knowledge were at the forefront of the company's priorities. In addition, during the interview my personal abilities and skills such as communication and team working were also assessed. I think the claims about difficulties in making the transition from academia to industry are, in my opinion, largely mythical. My academic expertise and ability to work on new projects were fundamental to securing my position. In addition, the scientists with whom I work are first rate and I have enjoyed interacting with them and learning about wider scientific areas outside my own expertise.

Career factors

There are many opportunities to get into industry at entry level and one, possibly two, postdocs is sufficient to make the transition. If you are a more senior postdoc you can use your network to convince potential employers of what you could bring to their company in terms of your expertise, skill-set and ability to handle a lot of projects.

Commentary: think 'skills' not research discipline

Having started his PhD in chemistry with a large biological component, Andrew's postdoc work had been focused on enzyme design in bacterial systems. In applying for a position in an agricultural biotech company, Andrew saw that he could transfer his technical expertise and specialist knowledge into industry. Employers will, in the main, be more interested in your skills than in your disciplinary field and there is always scope for you to offer your research and technical expertise. It's all about pitching yourself in the right way and highlighting the areas which will be of most relevance to the post. If you see yourself in a positive light, you are more likely to win over the employer.

5. *John*: scientific team leader, drug metabolism and pharmacokinetics (DMPK), global contract research organisation

Career facts

Job description

Contract research organisations (CROs) conduct research work outsourced by pharmaceutical and medical companies. This includes screening, non-clinical testing, toxicology studies, safety pharmacology and metabolism, right through to clinical studies in humans with supporting regulatory affairs and final approval.

As the scientific team leader, I work in a 'matrix environment'. That is, we operate in pools of expertise from which we source people internally who will fulfil roles along the drug pipeline. Amongst other roles, I act as the project manager on early development metabolism and clinical studies, and lead the matrix pool of other study directors within the company. For my own DMPK study group, I design and manage the non-clinical and clinical studies associated with the metabolism aspects of the drug development pipeline. I will meet with the client to agree the work required, after which the study is designed including an outline budget and time-frame. The study director then drafts the protocol which is carried through by the operational staff (usually graduate bioscientists). We trace the radiolabelled form of the compound of interest (e.g. potential anti-HIV, anti-cancer, diabetes drugs, etc.) in animal and human subjects, following its fate. Excreta samples are analysed to determine routes and rates of excretion, as well as blood and plasma samples for pharmacokinetic studies. The structural identities of metabolites are elucidated through metabolite profiling, including mass spectrometry.

As well as running my own research study group, my role is focused on business development, with my primary responsibility being to look after our clients (new, old, current, lapsed, etc.). I need to generate new business and identify new clients, primarily in Europe but also further afield in the US and Japan. This involves visiting and hosting clients, liaising with heads of department and commercial groups. In addition, we organise scientific symposia and make presentations at conferences, and are at the forefront of driving new regulatory guidance for the work we conduct.

Background

Following my PhD in cell biology, I progressed to a 3-year postdoctoral position at the same university looking at the effect of statins on cholesterol homeostasis, which led me into the field of drug metabolism. However, looking ahead, I could see that the metabolism department was moving in a different direction so that within 5–8 years I would need to move on. Therefore, I decided it would be better to move sooner rather than later so that I could secure more stable employment early in my career. I was not relishing the prospect of having to move on every 3 years in academia and I knew that tenured positions were few and far between. I applied for quite a lot of jobs in the first instance. Big pharma was booming at this time, but I was more interested in working for a CRO as I considered them to be more flexible and adaptable to changes.

I was offered my first job in a large CRO where I worked as a senior scientist for 5 years. During the last 18 months of this job I learned the responsibilities of a study director which enabled me to apply for a senior study director post at a small CRO, where I ended up as head of the laboratory and study director group. From there, I moved into my current employment within a large global company which, although a lower position than my previous one, I considered would place me in a more secure position. I have since moved through grades at my current company and now head the study director group responsible for all *in vivo* preclinical and clinical metabolism studies.

Career factors

Match up your personal preferences

My decision to move into industry was not easy at the beginning. I knew that the working environment would be very different to that of a university; I would be more accountable and be working in a bigger group, I would probably have to start at a lower level and would not know anyone. However, my desire for stability and a more secure employment structure prompted me to move out of academia after my first postdoc, as I envisaged this would probably be inevitable in the long run anyway.

Skills and attitude

The interview process for getting into industry is quite different than for postdoctoral posts so you need to be well prepared. Don't be concerned about 'commercial awareness' as employers will be more interested in your laboratory and personal skills. Every pharma company operates Good Laboratory Practice (GLP), adhering to legal codes of practice. Therefore you will need to demonstrate accurate recording of data and excellent chain of custody practices (the chronological documentation of information). In addition, your ability to work in a team and to be flexible will be tested. Most pharma and other large companies operate a 'targeted selection' process which means they will be looking for your ability to apply your skills and knowledge into particular situations. They don't want people who are rigid and wedded to one specific aspect of research. They need senior scientists who have a positive can-do attitude who can adapt and have the potential for leadership. Interview questions will test applicants' initiative and positive outlook (see Chapter 7 for more advice on interview technique).

Commentary

John made the decision early on in his career to move into industry, even though he knew it would not be easy to begin with. He was used to the university environment, having studied and worked within it since he was an undergraduate student. However, although moving away from his 'comfort zone' was a bit daunting, John decided to make the transition early since he was sure that the academic career path would not suit his desire for more stable and secure employment. Transition and change are not easy in many situations in life (see Chapter 8, Box 8.1), but sometimes facing up to them sooner rather than later can make the process easier. In many cases you can test out potential new career paths by doing some work shadowing or voluntary work during your research post. For example, if your group is collaborating with an industrial/business partner, you may be able to arrange a lab visit.

6. *Joanna*: product and R&D manager, small biotech company

Career facts

Job description

When you work for a small company, it is important to be flexible and to be able to take on a number of diverse roles to keep the company developing, so it retains an advantage in the market. A small company faces greater risks than a larger one so a versatile attitude is essential. I was taken on in order to utilise my expert knowledge in plant science, which I gained during the course of my research. I have been working with the company for 10 years now, during which time I have identified and developed a range of plant antibodies and innovative automated instrumentation. Although I have many roles to play within the company, it is communicating with researchers as well as coming up with new ideas and being able to see them through to a final product which I enjoy most.

Background

My personal circumstances have very much influenced where my career has taken me since my only reason for moving to Sweden was because of my Swedish partner (now my husband). We had met in Poland during the final year of my PhD when I was working on heat-shock proteins. I then secured two postdoctoral positions, the second one being in Sweden, which suited us both for work and personal reasons. As my postdoctoral position came to an end, I applied for a research grant which was rejected and I knew this was the turning point in my career; I made the decision to leave research and move into industry. It seemed an easy decision at the time – I had started to become slightly frustrated with research and was thinking of a career where I could still use my plant science training, but where I would see more immediate and tangible results with obvious utility and application. I started a Swedish language course to increase my employability as well as writing around 10–12 companies to find out what they were doing as part of my job search. As it happened, my present company was considering employing an expert to help develop the company so my email arrived with perfect timing and I was offered a job which has since become my own, i.e. I have been able to steer the job and take ownership of it, even acting as the design and advertising manager due to a useful family hobby in photography, which has helped me to design our website and produce attractive posters and flyers. All in all, it seems that many of the aspects of my life so far were leading up to this job and it has enabled me to realise my dream career (even though I didn't realise it at the time!).

Career factors

Don't be afraid

Many people consider small companies to be a bit risky and target large ones instead. Whilst there's nothing wrong with working in a large company, it doesn't suit everyone. For me, I like autonomy and being able to see my ideas right through to the end-product. I also like the close-knit team in which I work and the sense that we are 'all in it together'. My advice to those thinking about small companies would be to search out those with around 5–30 employees (sometimes known as small or medium-sized enterprises, or SMEs), preferably with a forward-thinking boss. If you approach companies in the way I did, my advice is to come at it from the side rather than head-on; that is, enquire about the work of the company which gives you more scope for an opening rather than asking for a job directly, which could be turned down flat with no room to negotiate further.

Also, don't be afraid to show who you really are when you approach companies or when you attend interview. If you aren't your genuine self and pretend, for example, to be knowledgeable or capable of everything, you may come across as being false. The important thing is to have the right can-do attitude which will instil confidence into a company director, who will be relying on your input far more than a company of thousands.

Be flexible

Generally speaking, small companies cannot afford to employ people who can't be adaptable and turn their hand to a range of tasks. That's not to say you would be expected to be able to do anything and everything; I would not be able to program a computer but I can offer extra input in marketing and advertising. Others may be able to assist with the technical or financial side, for example. Dedication and commitment are crucial but it is also important to ensure you don't end up working too hard, as I did at one point. Self-management is a useful skill to keep your sanity!

Commentary

Joanna knew what she wanted to do at the point in her career when her postdoctoral research post had come to an end and her grant application had been turned down. Turning points such as this and a realisation of the talents you possess and how to use them help you to search for alternative career options. Different people find different parts of their research interesting or frustrating according to their personality and the skills they enjoy using. This is a useful clue to the kinds of jobs which may interest you in the future. In addition, Joanna proactively targeted small companies during a creative job search which involved her making speculative enquiries rather than waiting for a job advert to emerge. The majority of jobs are not advertised (see Chapter 4) so this kind of proactive strategy can reap rewards.

In Joanna's case, personal circumstances dictated the direction of her career. Regarding your career as flexible and taking a creative approach to it will make this situation easier. As with all areas of life, compromise will mean that you can't always get what you want in one part of your life, which may mean either making a sacrifice in your personal life (for example, by living apart during weekdays) or changing career direction.

7. *Petra*: scientific adviser, protein interaction services company

Career facts

Job description

In my role as scientific adviser for customer projects I am primarily office based and offer advice to our customers face to face, over the phone and by email on the best strategic approaches to take for their protein–protein interaction projects. Customers are based in industry, government or academia conducting basic or applied research, ranging from plant science through to cancer science. Working in a small team of 3–4 people, we discuss results, interpret data and relay information between our lab technicians and research scientists. I handle 30–50 projects at any one time, which are at different levels and stages of progression. Therefore, it is essential to be organised and to remain calm under pressure, as well as being able to multi-task.

Background

Following my PhD in molecular biochemistry on the technical development of protein–protein interactions in Lausanne, I knew I wanted to continue on to a postdoctoral position. In Switzerland you can apply for a 1-year fellowship when you complete your PhD so I contacted one of our research collaborators in Paris. I knew he was hiring postdocs to run his proteomics operation and I wanted to change to the more applied end of my field. Although this was not my specific area of expertise, I convinced the group at interview and during my presentation that I would be able to contribute fully to their research programme. The project extended a further year, during which time I decided I wanted to move into industry as I was not confident that the academic set-up would suit me. I applied for a number of jobs by targeting company websites – some were advertising posts and others I wrote to speculatively (in this case I looked up the company on PubMed Central to see who was publishing and I directed my CV to the last author who is, by convention, the lead scientist). Although my applications yielded no results to start with, some months later at the end of my postdoc, I received a phone call from one of the companies who was starting to hire but not in research, in the service sector. I had not thought of this side of business but, in fact, it turned out to be exactly right for me. Having made a presentation and been interviewed (in a very similar way to my postdoc interview), I was offered the job and started 2 weeks later.

Career factors

Your work is never finished in service jobs like mine as there are always projects in various stages of completion. However, you no longer have to experience experiments that go wrong! In addition, you don't have ownership of projects as you work in a team and are providing a service rather than conducting new research. It's fun having different things to do and I enjoy the variety and versatility of my job. I have contact with lots of people and also attend 3–4 conferences per year to promote the company to delegates. The hours are more regular, although I do put in longer hours from time to time if required. It's interesting that having once been afraid of leaving the lab, I am not missing it at all. In any case, even if you pursue an academic career you are nearly always bound to leave the lab eventually as you progress.

Commentary

Petra started her job seeking around 3–4 months before completion of her postdoc as she was certain she would like to move into industry. She did not have any contacts but used company websites to make applications, applied for advertised positions and was also very creative in finding relevant scientists to target speculative applications. Although she had been intending to pursue research, when a service position came up Petra went for interview anyway. As a result, she realised this career area was more suited to her than research, as it involved more variety and contact with people. It's worth testing out alternative options sometimes, as they may lead you to a more fulfilling career than the one you were originally considering.

8. *Ann*: president and chief executive officer, small immunoreagent company

Career facts

Job description

We are a small company with five full-time employees, four being scientists and one accounting/billing. Since we are so small, we all do almost everything in the company, from washing dishes to making product. The vice president and I are responsible for most of the customer contact and filing of product. We are also primarily responsible for deciding what path the company will take. Ultimately, it is my decision what we will or will not do, what we will charge for our products, who we contract with when necessary, and who we hire/fire. I take care of most of the business decisions and lead the scientific team.

I like the variety and not doing the same thing over and over. I really like the science and don't want to be in a position where I am removed from it. However, I enjoy learning about the business end as well, and am now learning about sales. As part of our commitment to give back to the community, we bring in high school and college students to do internships in the lab. For high school students, it gives them some experience in the science field, and we hope to stimulate their interest in pursuing a science degree in college. For college students, we give them much needed practical experience in the lab sciences to make their job search more successful.

Background

I went to college and fell in love with science, particularly environmental and wildlife. I obtained my Master's degree in ichthyology and focused on cytogenetics in fishes. From there I went to a medical cytogenetics laboratory technician job. This was not as much fun as working in the field, but there weren't very many job opportunities for ichthyologists at that time. During this time, I became interested in chromosome structure and function and so decided to do a PhD in cell biology. Subsequently, my postdoc work involved development of monoclonal antibodies for use in cancer diagnostics, after which I obtained a position at a diagnostics company where I was in charge of antibody development and procurement. I was also responsible for antibody testing and quality control, as well as contract negotiation for purchasing antibodies from universities. It was here that I learned about cGMP and ISO Quality Standards.

I brought this information/knowledge with me to a small company in North Carolina where I was in charge of manufacturing polyclonal antibodies used in immunodiagnostics. Because it was a small company, my responsibilities grew daily; I handled contracts with customers, worked with the sales force, and handled all technical support and troubleshooting with the top 10 immunodiagnostic companies in the world. It was when the owner decided to sell the company that I took the step to start my own company.

Career factors

Grow with your career

A career lasts a lifetime. It is not a static thing but a living, growing thing. Grow with your career in whatever direction you want to take it, or whatever interesting paths open up before you. Some of your choices will be fantastic and you will learn all you can and enjoy it immensely. However, some choices will not work out as you had envisaged. That is OK, because you now know more about yourself and what is important to you. Don't be afraid to try something new, just be true to yourself. If you hate your job, find another – it's best for all involved. Don't be afraid to make a mistake, everyone makes mistakes.

Commentary

People who make the decision to go freelance, do consultancy work or, as in Ann's case, start their own company have almost always become experts in their field. They are highly knowledgeable about the business, have years of experience and an extensive network of contacts. On top of this, however, personality and disposition will determine whether you are suited to running your own company. You need tenacity, business acumen and a passion for what you are doing in order to be successful. In addition, you need to identify potential staff or partners who will complement your own skills, as well as being dedicated to the success of the business. Good leadership is not something you can learn easily, but if you think you have the right attitude to go out on your own, the rewards can be greater than working as an employee.

9. *Yfke*: senior medical writer, medical communications agency

Career facts

Job description

Being a medical writer is challenging but can be very rewarding. You need to be able to write well (quickly, succinctly and knowledgeably) in a variety of ways and using a range of media (e.g. PowerPoint slides, scientific publications, congress materials, promotional materials), as well as writing for different audiences (clinicians, patients, sales representatives and researchers). You need to be able to comprehend, in reasonable depth, the field in which you are working – mine is oncology – so it is intellectually challenging. Working on behalf of the pharmaceutical industry, I also travel quite a lot to medical conferences where I liaise with clinicians to discuss and finalise their presentations for industry-sponsored symposia. In addition, I prepare abstracts for medical congresses and assist clinicians in writing research papers for submission to clinical journals, which requires a strong grasp of clinical and research data as well, as ethical conduct in accordance with Good Publication Practice (independent guidelines to ensure that clinical trials sponsored by pharmaceutical companies are published in a responsible and ethical manner).

Interpersonal skills are probably more pertinent in this job than I have encountered previously in academia; being able to negotiate, listen to clients, interpret their requirements and provide a supportive role are integral to working in a medical communications agency – these are skills that are less intellectual but equally important to the job.

Background

Having completed my degree in natural sciences (zoology) as an EU student, I decided I had the best chance of securing funding for a PhD if I stayed there and applied for a bursary from the university's European Trust. This was to be the first of many proactive decisions I have made during my career which paid off in the short and long term. I've always thought that if you want something, you have to work for it. So during my PhD I started to write for the university newspaper science page, interviewing lecturers about their research and converting it into readable interesting text that would be understood by a wide audience. This sparked my desire to pursue a science writing career of some kind, although I wasn't sure which area I was interested in or how to forge a career path in this rather unwieldy field.

It was during the time I was writing up my PhD that I managed to convince my PhD supervisor to give me a month off so that I could take up a voluntary press officer position with the Society for Experimental Biology to cover their annual main meeting. This involved interviewing researchers who would be contributing to the conference and then submitting press releases to the international science media. The work was fast, furious and fun and served to confirm my career aspirations to be a science writer. During the conference I experienced a pivotal moment when I met the news and views editor of a learned journal. This led to me covering her maternity leave for a year, during which time I entered and won a science writing competition. All of this experience helped me secure an interview with my current company. The rest is history!

Career factors

Be proactive

Shortly after the end of my PhD, my husband's research position took us to a new part of the country, which prompted me to look for jobs in this area. I soon discovered many medical writing companies were based in the region so decided to look into it further as a career option. Through a personal contact from my earlier volunteering, I found a lead into the company I now work for. Following some writing tests and interviews, my medical writing career was launched. Thus, my career philosophy is very much centred on the view that if you help yourself, others will help you too, and that no-one owes you a living.

Do more

Even if you are determined on an academic career, try to do more than the basic requirements of your PhD or postdoctoral post. Universities, especially, offer a wide range of opportunities to get involved such as writing for the university paper/magazine, joining societies, clubs, sports teams, and doing other more high-level activities such as supervising, teaching and organising seminars and events in your department. This adds weight to your CV and helps to widen your circle of contacts.

Yfke had a relatively clear idea about her career plans early on during her PhD. However, her goal to secure a science writing job was quite flexible, and she acquired a range of different writing experiences during her academic studies. This put her in a good position later on when the medical writing position came up, even though her previous experience had not been specifically related to this area of science communication. It is a good strategy to keep your options open and flexible during your PhD and not to turn anything down even if it seems irrelevant at the time – almost everything you do will jigsaw together and play some part in your career in the end.

10. *Carol*: freelance science editor, writer and training consultant

Career facts

Job description

With three jobs running concurrently – science editing, writing and training – 'portfolio worker' would probably best describe my career right now; that is, doing a number of different jobs for several clients which, together, make up a full-time career. My primary job at the moment is as a staff editor with an open access journal, which takes up about half of my working hours. This includes selecting which submissions to send for peer review, then overseeing the progress of manuscripts through the evaluation process and, finally, contributing to the decision to accept or reject. Everything is handled online, which enables me to work from home (France) whilst liaising with the other staff based in the UK and San Francisco. After spending the major part of my early career in scientific research, I find one of the benefits of publishing and writing is that it is possible to have a more flexible working arrangement.

Background

Research

As with many scientists, much of my research career consisted of a series of short-term contracts involving changing labs, supervisors and countries. During my PhD, I specialised in muscle biochemistry and started to develop an interest in cell membranes. During this time I went to hear a lecture one day on the Golgi complex, which was to prove central to the early development of my career. The lecturer was from the newly established European Molecular Biology Laboratory (EMBL) in Heidelberg. I wrote to him during the final year of my PhD to enquire about post-doctoral positions. He suggested I apply for funding from various sources and, less than a year later, I was working in his lab (1983–1986), supported for the first 2 years by a postdoctoral fellowship from the Royal Society, with an extension of 1 year from EMBO. By this time I was 27 and looking towards the US as my next career move.

I successfully applied for a short-term NATO fellowship to fund a postdoctoral position in Baltimore, following which I secured funding for 2 more years as a junior investigator. At this stage I decided to try for an academic/group leader position back in the UK, but I realised very soon that I was unlikely to receive an offer. So I decided to stay in the US and moved to California for my third postdoc. At that time I felt quite pessimistic about being able to pursue a successful career in academic research – there had been few female role models to bolster my confidence either in the US or the UK; the women I knew in high-level positions were mostly childless and I definitely wanted to have a family.

Publishing and writing

When I was working at EMBL, I often helped colleagues whose native language was not English to correct the manuscripts they were preparing, and I also wrote some small articles for the *Trends* journal series which I enjoyed doing. I liked to have a broad overview of my field, so I decided that a career in science editing or writing might be the route for me. When I saw a publishing job advertised in Cambridge I didn't hesitate to apply. The job was editor of a new journal and the timing was perfect – it was one of the turning points in my career. I worked there for 6 years developing a large network of contacts among leaders in the field, which has proved invaluable to my subsequent portfolio career. The flexible nature of the work also enabled me to juggle family life and my career (by this time I had three children under the age of 4). Even so, I took a career break during this time when the flexible working practices altered in 1996 and, through my contacts, started freelance writing for a number of journals. It was a friend who spotted the advert for my next post and passed it my way, saying it was perfect for me – and so it proved. I got the job – a newly created post as 'information special-ist' at a research institute in Cambridge – which involved helping to write grants and research papers, doing some administration and a little teaching to support a professor.

Personal circumstances intervened again (my French husband secured a perma-nent research position in Toulouse) and so with a need to find work in rural France, I was once again faced with the prospect of taking a creative approach to my career! Making use of my contacts, I attended an EMBL alumni meeting on the off-chance of creating one or two opportunities. It paid off as I became one of the founding members of the European Life Sciences Organisation (ELSO) which aimed to support early career scientists. I was the editor of *The ELSO Gazette*, the organisation's online magazine, writing editorials, keeping up to date with the latest policy issues, commissioning and editing news articles and mini-reviews of European papers, as well as organising career development sessions. The ELSO finally disbanded in 2008 and I embarked on my portfolio career. Fortunately for me, and again through my network of contacts, it was at this point that I was offered work with an online journal, which brings me to the current point in my career – who knows what will happen next?

Career factors

Contacts and confidence

It is clear that flexibility is the key when it comes to a research career but if you get to the point when you think you're not going to realise your potential as a group leader, for whatever reason, it is time to bail out and seek other opportunities. The contacts I accumulated over the course of my career, both in research and pub-lishing, have been vital but never more so than when I decided to change career direction. It is important to build and nurture your network of contacts as people will nearly always try to assist you in some way with your career if they can.

My confidence levels took a dive when I was trying to secure a group leader position when I realised there were few women role models for me to aspire to in the 1980s. Today, however, women are more visible in science, there are women in science support networks and even funding opportunities to help women and men alike return to work after a career break (see Appendix 4). Making positive moves and always being on the look-out for opportunities will keep you employed even if it is in a rather disjointed way – my career may not have been straightforward but it has certainly not been dull!

Commentary

Carol's career may seem all too familiar to those postdocs who are striving for an academic post and having to move from contract to contract. Her advice to develop a network of contacts and to be constantly on the look-out for opportunities is crucial to your employability. This is not only good advice for researchers, it is good advice for any career since a 'job for life' is a concept consigned to the past in most countries nowadays. Being self-reliant and adaptable are the key components to a career, whether it is a portfolio career like Carol's or a full-time post. No matter how much you plan your career, circumstances (especially your personal life) are likely to change your plans so try to take an adaptable approach and be ready to take opportunities which come your way.

11. *Mary*: features editor, *Teaching Tools in Plant Biology* (learned journal)

Career facts

Job description

I write *Teaching Tools in Plant Biology*, which is a set of online materials to support university-level teaching of plant biology that is published in a scholarly journal. Each teaching tool includes a review article written for undergraduate students, a set of about 100 PowerPoint slides that incorporate basic and advanced materials, to be selected from by the instructor, and a Teaching Guide that summarises key points and includes discussion and exam questions. Creating these materials requires me to read a lot, summarise complex ideas, and select a few experiments to highlight so that students can build their understanding of the scientific process. I also find and create photos and images to illustrate the materials in the slides. I travel to conferences several times a year to keep up with the field and interact with the community of plant biologists, and I also run teaching workshops for postdocs to help them prepare for faculty positions.

Background

I have always enjoyed teaching and figuring out how to teach difficult subjects and concepts. I had 14 years' teaching experience working at a primarily undergraduate university (PUI) where I rose to the position of senior professor. Prior to this, I did my PhD in plant molecular biology in New York and then went on to a take up a postdoctoral research position in California. As with other research-intensive universities, I spent the majority of my time on research and associated activities.

However, I was inspired by my professor's teaching methods and his general positive attitude towards pedagogy which showed me how exciting and stimulating teaching can be. It was this enthusiasm and my desire to use wet labs, as well as my research record, which secured me my post as assistant professor. I had enjoyed research but I was keen to move into a PUI which would be more student centred.

Career factors

Networking

During my research and teaching career I have built up a large network of plant scientist contacts both at a professional and personal level. In particular, joining a learned society, the American Society of Plant Biologists (ASPB), early on in my career gave me access to a wider spectrum of people and opportunities associated with my work such as attending their meetings at a reduced registration cost, keeping up with general policy, research and educational news through their newsletter as well as volunteering to help with activities. Eventually I gained a position on their education committee and took up the post of chair for 4 years. The ASPB organises a great programme of educational events at their main meeting each year and I have met and got to know many university educationalists over the years, which is now proving invaluable in helping me with my current work developing teaching materials.

Happenstance

My career is partly the result of two major decision crossroads in my life. First, it was on a lonely, miserable day between Christmas and New Year during my post-doc that I was in the lab tending my plants when I happened to pick up a copy of *Science* and saw the advert for the post of professor. I knew it was the job for me! Second, circumstances had brought me to a crossroads in my personal life when I decided to remain in Glasgow, having travelled over from the States on a visiting professorship in 2008. I had probably reached the point in my life where I was looking for a new challenge in my career but I didn't want to do something I didn't enjoy. I knew I liked teaching but didn't want to teach in school. Having started to write a teaching textbook, I also realised I enjoyed writing. I was looking around at my options and happened, one day, to see that a journal of the ASPB was advertising for a features editor to develop their teaching resources. I felt something like an adrenalin rush when I saw it – it combined plant science undergraduate teaching and writing – the perfect job had found me!

Commentary

Mary's career path shows that being open to opportunities and getting involved in extra-career activities such as joining a learned society can place you in a preferential position to be able to take advantage of happenstance situations. Chance circumstances play a part in many people's careers but you will still have to demonstrate that you have the knowledge and skills to do the job. You also need the courage of your convictions to take advantage of these opportunities and move around, weighing up your current situation against what may be seen, at first sight, as a calculated risk. Your intuition or 'gut feeling' will usually show you the way!

12. *Linda*: teacher training co-ordinator, public research institute

Career facts

Job description

I run a continuing professional development (cpd) programme for teachers to bring science into schools. Teachers come to the centre where they do experiments and talk to scientists, after which they can take exercises back to their school to use with the school students. Although well developed in some other countries such as the USA and UK, this is a relatively new initiative in Germany. I am the first person to be employed to set up the programme in our institute. I work in a small team of science communicators, all of whom have a PhD background apart from the team leader who is a professional journalist and another member of the team who is a science writer.

It is exciting to be involved in the design and implementation of projects, which come to fruition very quickly (within a few weeks) compared with research, which is much more open ended. In addition, I am more connected with the management side of things than I had been previously in my postdoctoral role. All in all, I find my new role more fulfilling and enjoyable than I found research – even when experiments were going well, I never really felt the role of a principal investigator would be suited to me. In this job we show teachers how to make highly technical science interesting and exciting. I like to see them eager to return to their schools to disseminate the new knowledge and bring the 'wow' factor of modern science to the students.

Background

Originally from Sweden, I undertook my PhD in cell biology in Germany and then did my first postdoctoral post in the US. My PhD work had gone very well and continuing in research seemed to be the logical next step. For my second postdoc, I returned to Germany to the same university after which I switched to my current institute in Berlin. During this time, I got involved in outreach and teaching as much as possible. Throughout the course of my research career, I took opportunities to get involved in communicating science. I went into day care centres to do science experiments with children, gave tours of the lab for school classes and exhibited science at public events. At the same time, I was teaching on university degree courses and even creating new courses for the department.

Towards the end of my third postdoc contract, I finally decided that an academic career was not for me. I had already been having doubts but I eventually realised that I wanted to change career direction to find something more suited to me, which I would find enjoyable and rewarding. As luck would have it, I happened to see an advertisement in the paper one day where my own institute was advertising this teacher training post. Of course, I applied and, because of my research background and the outreach and teaching I had done, I was accepted for the post.

Career factors

Awareness

After my PhD I automatically continued on to do a postdoc and then another and then another without really considering whether this was the right career for me.

I never felt wholly comfortable in my research role and much preferred communicating science to broader audiences. It was only when I realised that I did not want to continue along an academic career path that I started to analyse myself and increase my self-awareness. I am more of a generalist, I like finding innovative ways to do things and I enjoy communicating science to wider audiences through teaching and public engagement. At the institutional level, my awareness was also lacking. I was quite surprised when I saw this role advertised in the local paper. I never knew much about other departments of the institute, especially administration. I think this is true of many research scientists. We are isolated in our laboratories and do not make any effort to find out about the wider activities going on. In my current role I work with researchers to link them with teachers and I make the effort to keep them informed of what we are doing and the wider aspects of the institute to help broaden their horizons.

Commentary

Linda finally found her ideal job through what she calls a fantastic coincidence since the job was advertised at exactly the right time for her. These types of chance happenings are frequently cited by people, who say they happened to be in the right place at the right time. And whilst this is true up to a point, if Linda had not had the communication experience from all the voluntary teaching and public engagement she had done previously, her application would not have been successful. For most people who get their jobs through 'planned happenstance' (see Chapter 2), it is because they are already 'out there', doing things, talking to people, putting themselves in a position where a chance event may occur. Improving your self-awareness and awareness of your wider environment is fundamental to career planning; if you know who you are and the types of jobs which might suit you and your personal preferences, you will find that managing your career comes more easily (see Chapter 3).

13. *Ruth*: freelance science journalist

Career facts

Job description

As a freelance science journalist, I have the freedom to write on a whole range of science subjects and work for a variety of editors and publications. This includes producing articles of varying lengths and depth mainly for broad-based and specialist science journals and magazines. I was a bit worried when I first decided to go freelance as I was concerned about whether I would get enough work. However, there is a steady flow and plenty of ways to pick up work – freelance commissions are advertised on discussion lists and through science writer associations (see further information at the end of this appendix) and I use my network of contacts from previous work.

The irony with science journalism is that, generally speaking, the less you know about the subject on which you are writing, the easier it is to write the story. When you are ignorant of the topic you need to ask very basic questions to understand and make sense of the science under discussion. This makes you get right back to fundamentals, which helps you to write a story that is interesting and accessible to readers.

For me, working from home was a major impetus for going freelance but I appreciate it doesn't suit everyone. It's very flexible but you need to be able to balance work and home life (which means that although I'm at home, I still need to get a baby-sitter to look after my 6-month-old daughter so I can concentrate on the job). Sometimes you miss having colleagues around but the upside is that you don't have to engage with office politics. Here in New York there is a science writers' group that meets every now and then, so I can catch up and swap stories with my fellow freelancers.

Background

Following my PhD in genetics, I took up a postdoctoral post in stem cell research and it was at this time that I started to write science news articles in my spare time. I wasn't sure if research was the career for me so I was testing out a potential alternative. I had always had a penchant for writing ever since I was at school – I enjoyed it and it came relatively easily to me. I started writing news articles for a science website that published research updates written by scientists for other scientists. I wasn't paid for the articles, but they were fun to write and gave me valuable experience. I tended to write more when my experiments weren't working because it would cheer me up! I soon realised that, for me, science writing was more enjoyable and rewarding than research. In fact, when I think back now, I realise that much of my time during my PhD and postdoc was spent avoiding lab work!

I made my move into publishing by writing speculatively to an editor at one of the *Nature* journals and asking if I could go to their offices for a chat to find out more and to enquire about possible internships. He agreed to meet with me and this was a turning point in my career. Although there were no opportunities for me at that time, subsequently a 6-month locum position became available at *Nature Reviews Neuroscience*, covering someone's maternity leave. I took the job with a view to 'dipping my toe in the water' but I realised almost immediately I had made the right decision. My work consisted of commissioning, editing and writing, and I enjoyed it enormously. However, toward the end of my contract, a friend alerted me to a news editor vacancy in New York and I just had to apply. I was flown over for interview and got the job! I stayed here for a few years and little by little started taking on the odd freelance article in my spare time as I wanted to write about a wider range of subjects, not just cell biology.

Career factors

Write!

It goes without saying that if you want to be a science writer or journalist, you need to be a good writer. You may have plenty of experience writing for scientific audiences such as journal papers and reviews but writing news for wider audiences requires other journalistic skills. Take any opportunity to write: university newspaper, departmental newsletter, internet sites, learned society bulletins. Even writing a journal review will help to broaden your academic writing skills. You need to get practice and build up a portfolio to prove to yourself and to editors that you can write. Your first job may be an internship, placement or temporary position. This is a competitive industry and many writers and editors get in this way. If you prove yourself, employers will try to find you more work or a more permanent position.

Communicating variety

Working towards a PhD is the antithesis of being a science journalist. I spent 4.5 years focusing on a tiny subject working towards the production of a massive tome. Conversely, science journalism routinely involves writing 300 words in a few hours on any number of science subject areas. One day I'm writing about radiation maps from Fukushima, the next I'm relating a story on brain scans of people taking a hallucinogen. The variety is one of the aspects I love most about this job.

Once you have written and submitted your latest story, you need to be able to let go of your last piece and turn your attention to a completely different subject. You will likely have different news articles in preparation with various deadlines, word lengths and audiences. You need to be resourceful, have great time management skills and be motivated by deadlines. These can be harsh and adrenalin inducing, but if this prospect excites you, then this could be the career for you.

Commentary

The most common route from science to science journalism is to build up a portfolio of articles which demonstrates that you can write. Academic writing is a start but you will need to get a broader set of writing skills to prove your potential. Science journalism skills are similar to those required for writing good press releases (based on which, journalists write their stories). You can familiarise yourself with this style of writing by attending media workshops during the course of your research and reading general interest science articles such as those published on websites or in magazines. Internships and short placements are a great way to get short-term experience of the media and science communication. As well as your writing ability, your personality and disposition will determine whether this career is suited to you. You will need to be someone who can switch on and off quickly from one project to another, who likes working alone and who thrives on deadlines (see Chapter 3). Ruth went freelance after she had built up her reputation and contacts writing for reputable journals and magazines. However, as she says, the lifestyle does not suit everyone.

14. *David*: research associate, Committee on Science, Engineering, and Public Policy, National Academy of Sciences; Outreach Committee Co-chair, AAAS Science and Human Rights Coalition

Career facts

Job description

I am currently working on a study investigating the state of the postdoctoral experience for scientists and engineers in the United States, having recently completed the Christine Mirzayan Science and Technology Policy Fellowship (http://nas.edu/policy fellows), also at the National Academy of Sciences. In order to complete the study I have been taken on as a part-time member of staff until the end of March 2012. The rest of my time is spent applying for full-time policy posts, including an application for the AAAS Science and Technology Policy Fellowship. If my application is successful I will start the fellowship in September 2012.

My main responsibilities currently include collecting briefing materials and inviting experts to inform the committee members on topics related to the study. The study is an update to the 2000 publication, *Enhancing the Postdoctoral Experience for Scientists and Engineers*, also from the National Academies. Funders and other stakeholders have been invited to contribute to the study. We are also making comparisons with the experiences of postdoctoral researchers outside the US, which has thrown up some interesting findings, e.g. >50% of postdoctoral researchers in both the US and UK are international. The results of the study are due to be published in 2013.

Background

I realised during my postdoctoral research post that I wanted to transition from research into science policy. I had a research contract at a Scottish university, during which time I got involved in the university's postdoctoral association, ending up as the co-chair by the end of 2008. As the name of my local postdoctoral association became more widely known, I was invited as a keynote speaker to the Vitae research staff conference in November 2009, where I issued a call to action that highlighted the need for a UK-wide postdoctoral association. Once established, I was then voted in as the co-chair of the fledgling UK Research Staff Association (UKRSA). I later became more involved in national research policy and helped to set up a regional association for Northern Ireland and Scotland. All of this policy work was done on a voluntary basis whilst I was employed as a postdoctoral researcher, which meant working during the day on my research and then spending my evenings on policy – it was like having a second job!

My research contract in Scotland expired at the end of 2010, so I returned to the US aiming to change career paths into science policy. I figured I had sufficient savings to allow me to job-seek full time for several months, after which I would take a part-time job in order to continue with the process. I was aware of US science policy fellowships such as the Mirzayan Fellowship, that I recently completed. Also, the AAAS Science and Technology Policy Fellowship (http://fellowships.aaas.org), which offers a full salary and places you in executive branch agencies such as the Departments of Energy, Homeland Security, and Health and Human Services or Congress. During my job search, networking was my main strategy; I attended as many meetings as I could, including the 2011 AAAS Annual Meeting, the 2011 National Postdoctoral Association Annual Meeting and the 2011 AAAS Science and Technology Policy Forum. During the latter meeting I met with a delegate whose group I had applied to at the National Academy of Sciences, which gave me the opportunity to indicate my interest in his group.

Career factors

Take a strategic approach

If you know the career path you want to follow, as I did, it is important to be strategic. On my return to the US, I had already been involved in policy work on a voluntary basis and needed to extend this further to put me in a position to be eligible for more substantive policy fellowships or for full-time posts. My networking brought me into contact with people relevant to my chosen policy career, giving me an advantage when I put in my fellowship application. Now, with my additional experience, I think I am in a strong position to realise my original goal of becoming a science policy professional.

Communication skills

Policy work involves being comfortable and proficient when engaging with people either face to face or in writing. In addition, you need to be able to comprehend and write sometimes dense and complex policy information, extracting the relevant messages from multiple sources and synthesising them in a brief, useful summary. Technical writing experience is a start, but you need to extend your capabilities through voluntary work and extensive practice. When networking strategically, it's important that you are able to keep the conversation flowing and on track to make it useful and productive for both you and your conversation partner. Most people you speak to will have a business card so make sure you have one handy to pass around. Having a presence on LinkedIn (see Appendix 2) will enable people to find out more about you as well as enabling you to make contact with a wider network of policy professionals.

Commentary

If you are interested in science policy work, bear in mind that most jobs will be based in the capital city of a country so you need to be prepared to work there or in other large cities during the course of your career. There is no obvious entry point into this career so you need to take advantage of opportunities available during your doctoral or postdoctoral research post and network intensely. Become familiar with local, regional or national policy organisations first – most are under-resourced and will welcome an extra pair of hands. Once you gain experience and become part of policy networks, you can add this to your CV/resumé and use it to apply for more established positions.

David has created a collaborative Google Docs list of policy fellowships and internships and invites your contribution: http://bit.ly/A3bq0X.

15. *Edward*: clinical trial co-ordinator, university cancer trial centre

Career facts

Job description

As a trial co-ordinator, I project manage a range of clinical trials and oversee all the stages along the way. The clinical trials, which are based in hospitals, are quite lengthy and last an average of 5 or 6 years depending on the nature of the trial, the rarity of the disease, number of patients, etc. There is a lot of complex administrative and regulatory work involved with running a clinical trial, including applying for regulatory approvals from the government regulatory agency and ethics committees, preparing documentation and overseeing the running of the trial itself. The data generated from the trials must be carefully and stringently collected and analysed using databases and other computer management systems. I am also responsible for setting up and activating the trials which involves close liaison with the clinical trials co-ordinators based in hospitals, the chief doctor in charge of the trial and research nurses.

Located in the university haematology department, I deal mainly with transplantation trials. My counterparts work on other portfolios such as lymphoma and leukaemia but within the university a whole range of cancer and non-cancer trials is conducted. Many of our trials are initiated by doctors who want to trial procedures or combinations of drugs which they have been using on a small scale in their own hospital with the objective of changing national or international practice. They may wish to trial a drug or procedure which is being conducted in another country. We can assist the doctor in applying for funding to run the trial as well as providing advice about clinical trial regulations and preparing essential documents associated with the procedure such as treatment plans.

Background

My PhD was in developmental biology, although a higher level qualification is not mandatory for this job. Having said that, most people who reach my level have usually worked their way up from the role of data manager, so having a PhD helped me to gain entry to this career at a higher level. During my PhD I realised I did not want to continue in academia as it felt too isolated for me. I was looking for a career where I could still stay at the sharp end of bioscience and use my science knowledge and skills but in a team setting. I started to browse academic job websites as I was coming towards the end of my PhD to try to generate ideas and see what non-academic jobs were being advertised in universities. I noted that the Good Clinical Practice (GCP) professional qualification seemed to be necessary for quite a lot of jobs I was interested in so I took this course online to help increase my employability. This definitely helped me when I started making applications and I quickly secured my current job working within a university environment, but in a healthcare-related career.

Career factors

Research the job market

I began to look for jobs before I was due to complete my PhD and even though I was not in a position to apply for them at the time, it gave me a good sense of the job market. This is a really effective way to do your job search, especially when one of your variables is fixed. I knew I wanted to stay in a university environment as I enjoyed the atmosphere and culture when I was doing my PhD and there were plenty of jobs to choose from. However, don't just rely on the skills you have developed during the course of your PhD to get you into a new career – you may still need to think about further specialisation which can give you an advantage if you know roughly the area you want to work in.

Match your skills to the job specification

It sounds really obvious but many applications are a tick-box exercise. You need to recite back to the employers exactly what they require by showing evidence that you have the skills and qualities they are looking for. You don't have to have directly relevant experience, you just have to show you have the potential to do the job. The kinds of skills and personal qualities required in this job are highly relevant to scientists. As well as having a good knowledge of bioscience, you need to be able to manage complex information, show excellent attention to detail and an accurate and methodical approach, and be able to multi-task and prioritise your work. There is a huge amount of legal and regulatory documentation associated with the work and so you need to be highly organised. You should be able to show evidence of all these skills from your research.

Commentary

Rather than having a definite career plan or goal, Edward spent time perusing job vacancies looking for inspiration and careers suited to his interest and skills. He had a definite intention to use his bioscience knowledge and expertise whilst transferring his skills into a different career, which would be more team orientated. This is a very effective way to search for jobs and discover other career options. The job descriptions give you valuable information about what is involved and the kinds of skills you will need. If you are aware of your skills, values and interests (see Chapter 3), you can identify which careers would be more suited to you. If you spot a gap in your skills, as Edward did, you can try to fill it by doing a course or enhancing your employability in others ways (Chapter 5).

16. *Miguel*: patent examiner, European Patent Office

Career facts

Job description

The minimum qualification for a patent examiner at the European Patent Office (EPO) is a university degree in the subject relevant to the role. However, in reality, most (if not all) of the 250 examiners working in the biotechnology cluster have a PhD and postdoctoral experience. The reason for this is that the job requires highly critical analytical skills and the ability to have meaningful in-depth scientific discussions with the patent applicants. The information and analysis are highly specific and you need to be able to deconstruct the details of the application to determine its value and contribution to the advancement of a particular technology. In some cases, the decisions we take can have serious economic consequences for the companies involved and they are watched closely by different social groups concerned with ethical or ecological aspects.

Only 28% of biotechnology patent applications are granted per year and the process, from start to finish, lasts several years. There are opportunities for team work, and all final decisions are taken in teams of three examiners, but most of our work is carried out individually. Each of us works on our own applications and we have almost complete freedom and flexibility to organise our work. I research databases of scientific publications and patents, examine all the information, analyse the contribution and build up a dossier. I write a report highlighting the strong and weak points of the application and give my opinion on the patentability of the invention. I then contrast these opinions with those of the applicant's through an exchange of letters until we reach an agreement or their application is either abandoned or rejected. Often it is necessary to meet with the applicants and their lawyers at a hearing with two other examiners to make the final decision and to close the case. All decisions must be reasoned and are based on a quite unique combination of scientific and legal arguments, which is the essence of this job.

In addition to scientific expertise and a highly critical analytical mind, language skills are required by the EPO. English, French and German are the three official languages. Since most applications in the biotechnology field are drafted in English, you need to have excellent knowledge of English and a good knowledge of French and German in order to apply. If your language skills are not good enough you will need to improve on them before you are accepted for the job. For full information see: www.epo.org/about-us/jobs/examiners/profile.html.

Background

Towards the end of my PhD in molecular genetics in Madrid, I started looking for postdoctoral positions. I wasn't convinced about aiming for a career in academia; during my PhD I had seen many highly talented researchers striving for this only to be unsuccessful due to strong competition for very few permanent posts. Also, the inherent instability of consecutive postdoc positions in different labs over a period of 6–10 years to build up a good CV was not appealing to me. Therefore I focused on research posts in industry and secured a position in a small start-up company in Germany. The job was ideal – it was a permanent post and I felt more valued than in academia. I led a small team and could delegate routine tasks so I could focus on reading, experimental design and analysing data. However, after 2 years it was clear the company was not really taking off, so I started to look at other options. As well as looking for research posts, I saw an advertisement for a patent examiner in *Nature* which caught my attention. I had never considered this as a career before. I had been on the other side of the process as a patent applicant but I could never understand a word of the legal aspects of the document. I decided to apply to see what happened and as I looked more closely at the job and working conditions, I became more interested. I was attracted by the prospect of being at the forefront of technological developments and being privy to the very latest breakthroughs. Also, now that I understand the legal side of the job, I enjoy it as much as the science!

Career factors

Examine jobs in detail

When I applied for the patent examiner post I didn't know much about this career. The job advert caught my attention because it was science related. However, when I looked at the details more closely I realised that the specifications were well suited to my skills and interests. I would recommend that any researchers looking at vacancies should not just focus on the specific subject knowledge and qualifications. You get a much better idea of the job when you look at the list of duties and responsibilities associated with it.

Commentary

This is a job for someone who is comfortable working with highly complex scientific information across the field of biotechnology and who has very well-developed analytical skills. Furthermore, it is quite an isolated role with people working individually on their own projects with limited interaction with others. This prospect will be appealing to those who enjoy great independence and flexibility but to some it may seem like a lonely job. This will depend on personal factors such as those described in Chapter 3. Your personality and the skills you enjoy using are more important when choosing your career than your specific subject discipline. When you examine job descriptions, look at the skills and personal attributes they require and don't dwell too much on specific subject knowledge (see Chapter 4).

17. *Florent*: scientific officer, Marie Curie Actions, European Commission

Career facts

Job description
My job is a specialised scientific administrative post. I deal mainly with the Marie Curie Actions (http://cordis.europa.eu/fp7/mariecurieactions/home_en.html), participating in the lifetime of the project from beginning to end. When each work programme is established we launch the call for applications and then evaluate the projects submitted (I specialise in the life sciences). This involves calling in experts, ranking projects against a set of criteria, and subsequently negotiating contracts with the successful applicants, reviewing and agreeing the detailed content. When the project starts, I check the annual and mid-term reports, which requires me to make a site visit to meet and interview the project partners. My role is highly diverse and requires a range of administrative skills: excellent writing, comprehension and analytical skills; attention to detail and organisational skills all complemented by my specialist scientific knowledge and understanding of academic research. It's a job I enjoy greatly and if you can imagine yourself relishing this kind of career too, bear in mind that a good knowledge of at least two languages is also fundamental to the job.

Background
Languages have always come quite easily to me – being a French speaker and having learned English and Spanish at school, I increased my language skills further during an Erasmus year in Florence, Italy. Subsequently, during my postdoctoral research, I spent 2 years in Barcelona where I practised my Spanish. It's not necessary to have more than two languages to enter the European Commission but it is expected (for promotion) that you will continue to expand your languages to three or more (I have been learning German recently as well).

Following my PhD in molecular cell biology in Brussels, I continued in research with a postdoctoral research post in neuroscience and behavioural science in Barcelona. Right from the start of my postdoctoral position, I was applying for other jobs, including management positions in pharma companies. I realised at that time that I didn't want to pursue a career in academic research. I had always been interested in research policies and this, coupled with my language skills, proved a good basis for the initial application to the European Commission. It was a long process, lasting 18 months with each application process taking place every 6 months (happily it has been shortened down to a year since that time).

Career factors

Personal development
As a 'fonctionnaire' (having a permanent post) in the European Commission, I am expected to move around internally and also to undertake professional development

courses. This is something I value greatly since I believe that personal mobility, adaptability and a sense of moving forward create a stimulating environment and keep people motivated. One part of my job is to speak to PhD fellows about their careers and I always tell them to review the options open to them (not just academia) and to keep an open mind. When you're doing your PhD you're not always aware of what's going on in the outside world so it's important to make some kind of connection even if it is just keeping an eye on the job market. In addition, you can add value to your research post by taking on more responsibilities to develop your personal skills. I asked for supervision responsibilities during my PhD and also brought together other scientists to form a collaborative team during my postdoctoral position.

Commentary

Florent didn't plan his career but he has been purposeful in his approach. For example, he wasn't initially aware that he would need languages or that moving around during his research career would act in his favour for this particular post but his desire to tackle new skills and new situations showed him to be adaptable and capable of moving forward. The repertoire of personal experience and skills he had built up in addition to his fundamental research skills made him employable in a range of different sectors offering him similar diversity and intellectual stimulation. Reviewing the job market and keeping your options open is an excellent way to find out about the kinds of skills and experience employers are looking for. You can discover which jobs would be of interest to you and identify gaps in your skills and experience which you can try to address. Whatever you do, personal development will help you to move your career forward and increase your range of options.

18. *John*: healthcare analyst, self-employed partnership

Career facts

Job description

I recently went into partnership with a former colleague, having enjoyed a 15-year career with a global investment bank. A healthcare analyst is, essentially, someone who assesses the merits of making an investment in a given healthcare company. It involves applying your scientific and financial know-how and being well paid for it. You have to be able to get complex ideas/issues across in a very simple way and no two days are ever the same. I have attended all the major scientific conferences to listen to and meet with world experts in order to take a view on disease areas and, therefore, the value of a particular drug to a company. I made the move to go self-employed to spend more time with my young family. Having been used to working in a large firm with all the technical, administrative and training support you could ever need, self-employment offered its own unique new set of challenges, not to mention trying to ensure a regular income.

Background

My original career plan was to work in research within the pharmaceutical industry. I had always wanted to do research work which was applied; for me, academic research was often too theoretical and too far removed from a clinical endpoint.

Also, I was attracted by the prospect of a more structured career path. I applied for and secured a research post in a multinational pharma company during the third year of my PhD (in biochemistry) which motivated me to complete on time.

It is difficult to generalise about drug discovery and pharmaceutical research as it can be so varied. I was perhaps less prepared for the day-to-day repetition when working in an area of advancing science, whilst at same time looking, as a team, to generate drug candidates. In addition, when working in a given disease area/project team, the focus is implicitly narrow. This makes your job quite precarious – as I found out when I was made redundant along with many of my research group!

Seeing this as an opportunity to change my career path, I became interested in the role of healthcare analyst as I wanted to stay close to major happenings/news flow in industry, but to work in (what I saw as) a more exciting job. However, there were no available jobs at the time (despite me offering to work for free for a while to get experience). So, instead I applied for a job as a medical journalist to gain a broader knowledge of the pharma industry and drug pipelines. My transferable skill-set of a strong basic scientific understanding, good communication skills and working to deadlines gave me a distinct advantage over colleagues without scientific training. With this additional experience and, having learned and updated my numeracy and accounting knowledge and skills, I finally secured a job as healthcare analyst within an international global investment bank.

Career factors

Skills, skills, skills

We are told today, anecdotally, that by 60 years of age, the average person will have had at least three different careers. Thinking in terms of your skill-set rather than specific job responsibilities will make career change simpler. Trained scientists are arguably better placed than many in other disciplines to embark upon a wide range of careers with skills such as hypothesis forming and testing, attention to detail, an analytical mind, proven oral skills and the ability to understand highly technical and complex issues. All these make a scientist a very attractive job candidate, even where the role in question may not appear to be obviously directly relevant.

Industry is far less concerned about the actual research work you have carried out than your personality, skills and general aptitude/willingness to learn. In particular, you need to work as part of a team (including often with younger and less qualified team members who know much more about the specific area of research than you may do at first!). In the investment sector the same is true. It was clear at interview that scientists were looked upon as having very transferable skills: highly numerate (very important in an analytical role such as this), naturally sceptical and analytical, understanding of key drugs and their modes of action and thus able to assess a company's future pipeline potential, ability to work to deadlines and good communication skills (oral and written).

Network, network, network

My only major regret during my career is that I found out too late the importance of networking and never did enough. The old adage is truer today than ever; it is 'who' rather than 'what' you know which is often the critical factor in getting that ideal job. Networking is absolutely essential to your career even during times of stability. You never know what's around the corner. Postdoctoral researchers may think they are badly served by short-term contracts but jobs are vulnerable in any industry. Although I eventually achieved the healthcare analyst career I had been striving for, I may have expedited the process a lot more quickly and easily if I had been better connected.

Commentary

John's career has been relatively goal orientated. First, he undertook his PhD aiming for a research job in industry. Then, his time spent as a medical journalist and learning about finance was with the intention of securing the role as a healthcare analyst. However, despite having a firm career goal, you never know how things will turn out and alter your situation. John's redundancy meant having to make some new decisions about his career. Finding it difficult to achieve his goal as a healthcare analyst, he was careful to choose an interim career which would enhance his chances further. He also added to his skill-set during this time.

As stated in Chapter 2, although it is important to be proactive and have a career aim, you will always need to be ready for changes outside your control. This is a very good example of where a combination of structure and flexibility can enable you to remain focused, but 'peripheral vision' makes you aware of factors beyond the core of your central aim. Networking is definitely one of these factors, as well as persistence and a positive attitude. And, as John says, scientists possess a highly marketable set of technical and personal skills which will always be of interest to employers.

19. *Annie*: technology consultant, international technology consultancy

Career facts

Job description

My role as a technology consultant is to manage discrete technological projects in a range of businesses. I work in a close-knit team of six on my current project which will run for about six months. This involves working with the sales and marketing department of a pharmaceutical company on the development of digital techniques to enhance their activity and increase productivity in this area of the company. The consultancy offers its expertise to businesses worldwide and has built up relationships with many, which means that we are assigned a variety of projects. I have only been in this job for six months and so this is my first assignment. I was particularly keen to work in a life sciences company because of my background and interest in biosciences (in particular my subject discipline – pharmacology). The way in which you are assigned to projects is either to source them yourself through contacts you know in the company or the human resources department will work with you to find a role.

Technology consultancy is about enabling companies to use new technology or to better use the technology they already have. Each project lasts for a finite length of time and once it is complete, you move on to a new project. The company is very large and operates globally so we have thousands of projects going on at any one time. Technology consultants implement technology solutions. In my role I sometimes act as the conduit between the technology experts and the clients, so it's important that I can convert complex information into a language which is clear and coherent for non-experts to understand.

Background

I have always been good at sciences, which is why I took them at degree level and then progressed on to do a PhD in pharmacology. However, members of my family worked as accountants and I have always been interested in business and how the economy works. During my PhD I participated in the Biotechnology YES (www.biotechnologyyes.co.uk) scheme which aims to enhance researchers' entrepreneurial and enterprise skills. I thoroughly enjoyed the experience and was prompted to consider business and management as a possible career move. I applied for and was offered a job with an accountancy firm directly after my PhD, but I decided I wanted to stay close to science so took up a post with a learned society instead. However, although this work was team orientated and involved managing projects and working with others, I still felt I wanted to be part of a larger organisation where there was a more clearly defined structure and career path. In addition, I felt that it would give me a broad insight into how companies work across a wide range of industries.

Therefore, I applied for another graduate training scheme. Technology consultancy appealed to me as I saw it would offer me the opportunity to work within companies allied to life sciences, which meant I could still use my interest in science within my job. Many companies that run graduate training schemes take on large numbers of student and doctoral graduates each year and throughout the year. The application process for this scheme (and most other graduate training schemes) involved a more complex process than for other company entry routes. First, I attended an interview which centred on my experiences and aspirations, and what I knew about the company. I also had to do some competency-based tests which included questions about my values. Having passed this stage of the process, I then attended a one-day assessment centre where we did team exercises (case study) and other tasks such as an in-tray exercise (prioritisation). These lengthy application processes are quite daunting but if you make use of your university careers service as I did, they can normally give you advice and even offer practice exercises.

All those accepted onto the graduate scheme were then sent on a 4-week training programme which consisted of learning about the company and its people, its ethos and how it works. This was really useful to show you the bigger picture of the company and how you fit into it. The training programme also gave us the opportunity to gel as a team, and to learn new skills and methodologies such as computing skills and project management.

Career factors

Qualifications, interests and values

I was always interested in science and business but I chose to do science as I was good at it and thought it would always be useful to me in any job I decided to do in the future. It was for a similar reason that I decided to do a PhD since I never thought it would hold me back and could only be a positive thing to do. People sometimes think that by doing a PhD, you will be too specialised to enter other professions but this is not true. In fact, the further removed you are from academia (where having a PhD is nothing special), the more value it has since it is not very common in other professions. That's not to say it means you get a higher salary – I am on the same salary as graduates who are much younger and less qualified than me. Nor does it automatically propel you to higher levels in the company. People are promoted on a meritocratic basis within most industries unless your PhD and expertise are especially significant to the company's business. In the main, you will need to prove yourself along with everyone else. However, your skills and personal maturity, developed during your PhD (managing projects, interacting with people at different levels of seniority, working in teams and making presentation), may prove advantageous.

If you are interested in moving into business, look into it first and ensure it will suit you, not only in terms of your interests but also in terms of your values. The working environment and ethos of large commercial companies are very different from public sector organisations. You are expected to work long hours and, in this business, working away from home for long periods of time may be necessary. However, the benefits are great and include a structured career path, good salary, pension benefits, subsidies such as gym membership and very favourable maternity leave.

Commentary

Having a PhD and research experience is transferable to a wide range of professions. Employers are interested in taking on scientists who are not afraid of technology and have other associated skills such as problem solving, analytical, numeracy and critical thinking. You may possess these skills, but you need to reflect on whether they are ones which you enjoy using and want to transfer into another profession. If they are, you will also need to consider other aspects such as the working environment, ethos and culture of the company. Many professions can be conducted in a range of companies and sectors, large and small, charitable, not for profit, etc. Working in a large company may suit you to start with, providing you with structured training and even a professional qualification which you can then transfer later on to a smaller business. Refer to Chapter 3 on self-awareness which provides some personal exercises to enhance your understanding of your personality, values and skills.

20. *Ian*: sound engineer, production services provider

Career facts

Job description
The company I work for provides lighting, audio and video equipment to the concert-touring industry. My job as sound engineer is split between sales and designing installations for clubs, sporting and conference venues. I also do live concerts with international rock bands touring for weeks and, sometimes, even months on end where I set up the sound equipment and mix the band as the 'front of house' sound engineer. This may seem glamorous on the surface, but the culture is far removed from the sex, drugs and rock and roll image of the past. Nowadays, it is a serious career and working hours can be from 8.00 am through to 1.00 am for 3 days in a row. What I like most about the job is that every day is different, each

one throwing up its own set of challenges and problems to be solved. The downside of the job is when a band wants very specific requirements, or when the weather is bad when we are doing an outdoor event. More recently, I have been appointed as the company's training co-ordinator so I will be running courses and teaching people how to use our equipment. This is ironic since it was one of the areas of my research which I enjoyed the most and now it has come full circle.

Background

I chose to leave my research career because I realised I was not passionate about the project on which I was working. I had decided to carry on doing a postdoctoral post after my PhD in order to stay in the local area but the project did not interest me. I also realised that my future was probably going to consist of short-term contracts and the prospect of moving every 3 years to find work did not appeal to me. During my PhD I had enjoyed my project and got on very well with my supervisor, but the postdoctoral post was very different. Most of the time was spent working on my own and I felt even more isolated as PhD students left and weren't replaced. On top of this, I found it frustrating that there was no real recognition for the work I was doing.

Music had always played a part in my life; I was in a band during my undergraduate degree and had started working part-time during my PhD for a local sound engineering company. I continued this work during my postdoctoral post with a larger company and when they offered me the opportunity to go on tour for a few months I took the decision to leave academic research. I loved it and never once looked back or had regrets. It ticked all the boxes for me: I was learning new things, new techniques, new protocols and getting positive feedback. The work involved close team working and I felt valued and appreciated. People listened and were interested in what I had to say.

Career factors

Knowledge versus skills

Sound engineering requires similar skills to those I was using for my PhD which was in biophysics. I had been amplifying signals, using recording instruments and employing my knowledge of maths and physics which I translated into sound engineering very easily. It was most likely this underlying interest which first attracted me to the PhD project. However, by the time I reached the postdoctoral stage other factors were having more influence on my enjoyment of the role so that the personal skills and working environment were not suited to my personal preferences. I like to solve problems and see rapid outcomes, work in a close-knit team, experience new challenges and variety. None of these were associated with my research work which is why I ultimately drifted away from it.

Don't be afraid to leave

I never regret doing my PhD. It gave me a great sense of achievement. But you really need to think hard about continuing along the academic career track – don't just get sucked into it like I did. A PhD can open a lot of doors to you, not just a postdoctoral position. If you are doing a postdoc now and are not enjoying it, seriously consider leaving. You never know what life has in store for you and now that I am teaching I can draw on the lecturing and mentoring experience I gained during my PhD and transfer it into this new and exciting role where no day is the same!

Ian demonstrates a clear example of where personal interests overtake academic or research interests leading to a career in areas such as the music business, mountaineering or as a professional photographer or ski instructor. Furthermore, Ian cited his need for closer team working and more immediate outcomes to give him a sense of job satisfaction, which he was not experiencing in his research career. Chapter 3 focuses on these personal factors which are more influential to your career choice than your academic knowledge. We tend to start our careers on the basis of our educational interests, knowledge and subject discipline but as our careers progress, we realise there is more to work and job satisfaction.

Conclusion

These 20 career narratives represent a tiny snapshot of people's careers and the factors which played a part in shaping them. The following websites provide further case studies and information charting the careers of doctoral and post-doctoral researchers.

Association of British Pharmaceutical Industries (ABPI). Case studies (industry). Available from: http://careers.abpi.org.uk/case-studies/.
Career publications and websites with individual career stories and interviews: www.vitae.ac.uk/researchers/1341/Career-stories.html.
icould stories. Includes interviews with people working in the pharmaceutical industry, health sector, environmental careers and many other non-science careers. Available from: http://icould.com/watch-career-stories/.
Vitae. (2004) *What Do PhDs Do?* Available from: http://ow.ly/7S6Km.
Vitae. (2010) *What Do Researchers Do?* CRAC. Available from: http://bit.ly/xb7MKM.

Social media

The digital age is set to alter the fundamental way we work and even the way we secure a job or livelihood in the future. In this global economy, digital technology is already enabling us to cross distant boundaries and enter new worlds to which, previously, we only had limited access. Therefore, it is absolutely essential as a working professional that you engage with digital technology or you will most certainly be left behind very rapidly. As researchers, you are already using electronic media such as email, digital repositories to store your data, discussion lists and the World Wide Web. Interactive digital media is increasingly becoming a normal part of everyday life. Many people write their own blogs or contribute to organisational blogsites. Jobs are being advertised on social media sites, so you may miss out in the future if you are not part of the right digital communities. Being able to demonstrate to employers your engagement with particular networks and specialised e-groups will show them you are up to date and interested in the wider world.

Digital media are still not an exclusive forum for conversation and communication – talking face to face informally over coffee or during breaks at conferences will remain the preferred style for many people. However, webinars, online live discussions and social media are starting to be used more frequently. This is just the tip of the iceberg so if you are not already doing so, get on board the social media train as soon as possible. Don't be left standing on an empty platform.

Blogs

If you are thinking of writing a blog, make sure you have something interesting to say and that you write at regular intervals. You may want to blog about your research, related issues or general scientific news associated with your research field. Perhaps you are interested in other areas of your life which you want to focus on. Bear in mind that you will be visible to the world, so make it professional and accurate so that it reflects you in the best light. If you make your blogs topical and engaging, relatively short and include photographs and images, you should gain a following quite quickly. You could even start a video blog (e.g. http://aobblog.com/).

Career Planning for Research Bioscientists, First Edition. Sarah Blackford.
© 2013 Sarah Blackford. Published 2013 by Blackwell Publishing Ltd.

You can increase your following and the number of people reading your blog by actively telling people about it using other digital media such as email, Twitter, Reddit and LinkedIn. You can share your new posts there and this will enable people to pass on or 'retweet' your links and also provide you with feedback. Google Analytics (www.google.com/analytics/) provides statistics about your blog, such as how many visitors you have or how they found your page. You can also consider including guest posts by other specialists.

You can set up your blog using any number of free hosting services such as www.wordpress.com, https://posterous.com/ and www.blogger.com.

Free/cheap images

There are a number of websites available that will enable you to liven up your blog or website. It is worth checking out the following.

- http://commons.wikimedia.org/wiki/Main_Page
- www.123rf.com/
- www.shutterstock.com/
- www.bioscience.heacademy.ac.uk/imagebank/default.aspx
- http://en.wikipedia.org/wiki/Wikipedia:Public_domain_image_resources
- www.cellimagelibrary.org/

See Appendix 4 for examples of bioblogs.

Twitter

I only recently joined Twitter as I had been under the impression that it was too intrusive and I would be constantly bombarded with messages all day long. With so many email messages arriving in my mail box already, I thought I would be overwhelmed. However, I was wrong! You can access your Twitter account how and when you like by choosing your settings so I only view mine when I am working on my computer.

Twitter is like a stream passing by which you can dip in and out of. With so many messages, you may miss a few (or a lot!) but you can organise them into lists and use various other ways to manage the flow of information.

What's great about Twitter is that it is much more open than other social media. You can easily find people of interest and significance to follow and then, by seeing who they follow, find further people to follow who may then follow you back. It's like a society in itself and once you've joined you'll find you enjoy being part of the Twitter community. In terms of your career planning and management, this is a great way to keep in touch with the latest news and developments in your areas of interest. Your Twitter account will demonstrate that you are engaging with the wider world and that you are motivated to keep abreast of issues, news, events and initiatives associated with a particular career sector or field of interest. More directly, jobs are advertised on Twitter and links to specialist blogsites can keep you up to date with career development resources, workshops and conferences.

Twitter makes use of shortened URLs so that you can fit your message into the 140-character limit. This can also be useful if you are using a long and complex

URL for an article or book (I have used some in the Resources section of this book). URL shortening sites include the following.

- https://bitly.com/
- http://tinyurl.com/
- http://ow.ly/url/shorten-url

See Appendix 4 for examples of bioscience career-related Twitter accounts.

Facebook

Many organisations have their own Facebook site, which is generally open to anyone who wishes to join them to be part of their community. You will find out about their latest initiatives and, in many cases, you can contribute to the site. For example, learned societies get very busy in the lead-up to their conferences or when they are advertising their latest funding initiatives. Learned journals also flag up recent or forthcoming published papers. Bear in mind that they also do this on Twitter, which might link to the Facebook site or their blog.

LinkedIn

LinkedIn contains features which are more suited for professional use rather than for personal and social interactions, such as Facebook. Therefore, for activities such as enhancing your career profile, networking and job seeking, I would recommend you set up a profile on LinkedIn. Employers, recruitment agencies and head-hunters use LinkedIn to identify potential candidates to approach. Joining specialist groups enables you to converse with people in job sectors of interest to you or you can simply 'lurk' on the sidelines to see what people are talking about. Careers advisers use LinkedIn to help people with their career strategy, job search and networking – there are ways in which you can find people to approach speculatively or enquire about work shadowing/volunteering.

LinkedIn profiles take longer to write than those for Twitter and Facebook so it is advisable to have plenty of information relating to your current and previous positions, education and specialist skills already available in a Word document so that you can transfer them easily into your new LinkedIn profile. This will ensure that it is as fully formed as possible so that when you start to contact people to ask to connect with them, they will be able to see a full profile rather than just your name and a few scant details. If you upload your CV you may wish to remove more personal information such as your home address and telephone number, but this is up to you. Try to include lots of keywords so that employers looking for people with particular skills will be able to find you easily.

These are just a few tips on how to use LinkedIn as part of your career planning strategy. Further useful insights into using LinkedIn to research industries and the professionals working within them and to search for jobs can be found in the following website tutorials.

- http://bit.ly/vRWir
- http://slidesha.re/aLyUkc

- www.slideshare.net/helenpownall (note that, although aimed at careers advisers helping students with their job search, this webinar presentation offers useful additional practical advice).

Social media snippets

Digital media are a great way to stay in touch with what's going on in your field and in potential career areas which you are considering. You can also pass on information you think is useful to others who, in turn, may do the same and so the network expands exponentially! Here are some snippets I have seen recently.

Newspaper article

Times Higher Education Supplement – 'No fixed address' (Paul Jump): http://bit.ly/jlGXQx

> … a steady rise in the number of postdoctoral researchers means the picture is very different today. Part of the increase is attributable to the fact that postdoctoral positions now proliferate far beyond their origins in the physical and life sciences. Today only half the posts are found in these disciplines. According to the US National Science Foundation's *Science and Engineering Indicators: 2010* report, the reasons for the swell in numbers are not well understood, however. It says that 'increases in competition for tenure-track academic research jobs, collaborative research in large teams, and needs for specialised training are possible factors'.

Article about securing a teaching post linked from www.jobs.phds.org and advertised on LinkedIn (Postdoc Forum group)

American Society for Cell Biology – 'How to get a teaching job at a primarily undergraduate university' (A. Malcolm Campbell): www.ascb.org/newsfiles/teaching.pdf

> If you know that teaching at a PUI is why you are getting a PhD in the first place, then you want to think about the implications of choosing a particular lab for your thesis work. If you choose a lab that does only one technique and your project requires you to work with live Ebola virus, then you are not setting yourself up for a teaching job. Most PUIs want a person who is versatile and can conduct student-based research.

Event advertised on LinkedIn (Postdoc Forum group)

I reposted this on relevant Facebook sites and the event diary of my blog and Tweeted it too.

Association for Women in Science – Central Arizona: http://awis-caz.org/

> AWIS-CAZ has a local graduate student club based at Arizona State University called Women in Science. This group works in association with the local AWIS chapter to develop enrichment seminars for graduate and

undergraduate students at ASU. Having a direct connection makes it possible for AWIS organizers to stay informed of the current needs and interests of the student population so we can better serve our community of young scientists.

Blog about postdocs, with associated Tweet advertising it to Twitter followers

Labguru – PI seeks postdoc for long-term relationship: http://blog.labguru.com/pi-seeking-postdoc-for-long-term-relationship/

In order to help graduate students home in on a successful and targeted postdoc search, Labguru asked several professors at elite academic institutions to provide feedback about what they look for in potential postdocs, what qualities impress them during interviews, and their advice for candidates. So before embarking on your postdoctoral interviews, read our insider tips that will put you ahead of the curve!

@labguru
Labguru is a new research management tool which helps researchers plan experiments, track progress and get results. Free for personal use. Powered by @BioData. http://labguru.com/'RT @wileyscience: great advice! RT **@Labguru @BioSoMaya**: PI seeks PostDoc for long term relationship http://blog.labguru.com/pi-seeking-pos…'

Blog about a science communication event, with associated Tweet

BioscienceCareers: www.biosciencecareers.org/2011/11/standing-up-for-science.html

Always say yes to a request from a journalist to talk to you. If you refuse they will probably write the story without you anyway, without your input. In the main, journalists are on your side and want to get the story right. If you build up a good working relationship with them, and get known to be reliable and accessible, they will come to you for comment on other related issues.

@BiosciCareer Sarah Blackford
'#voysmediaworkshop. Many thanks for a great day! See quick write-up here: http://ow.ly/7ypvV'

General scientific interest

@newscientist
New Scientist is the world's leading science and technology weekly www.newscientist.com 'The Smart Guide to 2012: Ten ideas you'll want to understand. www.newscientist.com/special/smart-guide-2012'

@jimalkhalili University of Surrey
Scientist, author and broadcaster
www.jimal-khalili.com
'Listening to @RogerHighfield on #TheLifeScientific talking about Colin Pillinger and Beagle mission #BBCRadio4'

Using social media: a personal perspective from Dr Anne Osterrieder, Oxford-Brookes University

If you are a PhD student or postdoc, the most obvious career strategy is to focus on your research and produce a lot of high-quality research publications. However, competition is fierce and in times of decreasing funding things will not get easier. To stand out from the crowd you should consider additional strategies to expand your network and make yourself known as an expert in your field.

Social media websites like Facebook, Twitter, Google+, YouTube or Reddit might not be the first thing to spring to mind when thinking about effective career tools. But if used right, they offer so much more than a way to procrastinate. Social media can help you to actively shape your online identity, which is the information people find when they look your name up in search engines – something more and more employers are doing nowadays. Using social media allows you to establish yourself in online networks that are not restricted to researchers from your own field but can also include professionals from related areas (such as science communication) or interested non-expert users.

Which social media platform should you use? This is entirely up to you. Some people prefer to post the same content to all of their networks, whereas others use different websites for different purposes. No matter whether you prefer to focus on one medium or register your name on all available social media websites, the important thing is to find a platform that makes you feel comfortable, both from a technical side and its user base.

I signed up for Twitter after I registered online for the British Science Association's Science Communication Conference and noticed that most of the attendees had a link to their Twitter account next to their names. Through the conference's official hashtag (#), I was able to contact other participants beforehand – which was great because I didn't know a single soul there and I felt less nervous when I arrived! Even more importantly, Twitter allowed me to keep in touch with people afterwards and to continue exchanging information, ideas and friendly chats. Furthermore, there is a whole 'blogosphere' out there that exists in parallel to traditional media and publicises itself online. Reading the thoughts of expert bloggers on topics such as current science stories, science communication, higher education or gender issues and other relevant topics has broadened my horizon more than I ever could have imagined.

Since then I have taken the plunge to start my own plant science blog. I use a number of social websites to share my posts and I have also set up LinkedIn to display my recent Tweets and blog posts. Every now and then someone I meet comments on how they enjoyed reading my links. Through this kind of feedback, through 'like' and '+1' buttons and through my website statistics, I can see that a lot of people are interested in online science and I can monitor the impact of my online engagement activities.

A lot of people tell me that they 'don't get' Twitter or other social media websites because they couldn't imagine that other people would be interested in their status updates. I think the most important function of social websites is not to broadcast yourself, although they are invaluable as tool to publicise your own activities (and believe me, a lot of people will be interested in what you write as scientist!). It is to share common interests with like-minded people, to expand your network and your horizon and to keep up to date with current issues. All in all, it's personal development – and much more fun than some official training opportunities!

@AnneOsterrieder

www.LinkedIn.com/pub/anne-osterrieder/35/670/341

Example CVs

Appendix 3 demonstrates how the content of your CV can be restructured and reformatted according to different vacancy specifications. An analysis of the CVs explains how each has been changed depending on the vacancy information.

Six example job advertisements are listed below, each of which describes the job role with its corresponding set of applicant criteria. Following this section are six corresponding CVs, which have been adapted to these specifications. The final section provides an analysis of each of the CVs and explains how they have been adapted to fit each of the job requirements.

Example Job advertisements

Job 1. Postdoctoral research scientist (research institute)

This project will investigate the role of chemotaxis in the spread of melanoma. This project will concentrate on the attractants that induce melanoma cells to migrate out of tumours, the mechanisms that generate gradients of these attractants, and the mechanics of how the cells respond. The work will involve culture of various melanoma cell lines, generation and transfection of GFP- and similarly tagged molecular markers, and high-level microscopy. Systems include high-resolution analysis of 2D chemotaxis in chambers, 3D and organotypic models of cellular invasion, and *in vivo* models of tumour metastasis.

A relevant PhD is essential. Experience in microscopy, cell biology, molecular biology and *in vivo* biology and a good record of quality publications would be an advantage. Applications for the above post should send their CV and names of two referees.

Job 2. Scientific officer (research institute)

The research aims of our laboratory are to identify therapeutically relevant mechanisms of drug resistance in patients with ovarian cancer using genomic and functional approaches. We are looking for a senior scientific officer to be responsible for supporting translational research projects based on primary ovarian tissues and ovarian cancer stem cells which will include but will not be limited to handling of primary human tissue, flow cytometry sorting of primary cells and

Career Planning for Research Bioscientists, First Edition. Sarah Blackford.
© 2013 Sarah Blackford. Published 2013 by Blackwell Publishing Ltd.

the establishment of new cell lines. Functional experimental work will include the detailed characterisation of primary cultures including drug screening assays and loss and gain of function experiments with viral vectors. This post will also support experiments in established cancer cell lines including targeted mutation to develop knock-out and knock-in models, drug screening assays and loss and gain of function experiments with viral vectors. Secondary duties include routine laboratory management tasks and training of new laboratory members. This post requires extensive knowledge of standard laboratory techniques including standard cell, molecular biology techniques, 2D and 3D tissue culture, cell-based sensitivity assays and assays on primary mouse or human cells. Proven experience in problem solving and improving established protocols will be essential and a PhD highly desirable. The successful applicant will be able to provide examples of being both a good team player and the ability to work independently. Excellent communication skills are essential.

Our online application process involves attaching your CV and covering letter which should outline your interest in applying for the role and explain how you meet the set criteria

Job 3. Communications manager (European science organisation)

We are seeking a communications manager for a European-wide science communication initiative funded by the European Commission. A network of communication officers from 10 European biomedical research institutes, we aim to improve the public announcement of health research to the general public and the media.

The communications manager will join the communications team, which is responsible for the international communication and outreach activities of the laboratory. He/she will project manage the media work package and work closely with the communication officers of the participating research institutes. The post holder will develop and implement media guidelines and act as a reference point for media work within the consortium. The post holder will need to travel and take part in project meetings.

Key responsibilities will include:

- compiling and writing a media guidelines handbook based on generally accepted guidelines in science communication and interactions with the press
- liaising and agreeing on these guidelines with the communication officers of the participating institutes
- developing strategies and consulting the communication and project officers of the participating institutes and projects to support their media work
- developing tools for collating and analysing media coverage
- liaising with international science journalists' associations and European media colleagues and organising media events
- delivering progress reports and presenting the project outcomes internationally.

Qualifications and experience
A university degree, experience in media work and an interest in life science research. Applicants must have proven project management skills and an ability to work within an international multicultural environment. Applicants should have excellent spoken and written English skills and advanced knowledge in other European languages would be an asset. Applicants should have strong

organisational and time management skills to manage conflicting priorities and to ensure work is planned, prioritised and delivered to meet key requirements and deadlines. As well as being able to work independently, candidates should have strong interpersonal skills, the ability to communicate clearly and effectively at all levels and be able to work effectively as a member of the international consortium as well as the communication team.

Job 4. Trainee chartered accountant (international accounting firm)

We are an international accounting firm and business advisory group, represented in over 50 countries worldwide. We offer a variety of services to a diverse client base, from large corporate organisations to owner-managed businesses across many market sectors, operating internationally. We recruit around 60 people each year to train to become chartered accountants. We'll support you through your exams, which usually takes 3 years. We are looking for candidates with a good academic ability and who are able to express themselves clearly both orally and in writing.

Key employability skills include:

- *flexibility* – to deal positively with changing priorities and workloads
- *curiosity* – to help you learn, share and innovate
- *team-working skills* – so you can contribute to and drive our success
- *commercial awareness* – so you're always up to speed with the latest business thinking and trends
- *great communication skills* – to help you listen, persuade and put your point across.

You can apply online by going to our website and completing an online application form.

Job 5. Assistant professor, nutrition department (university)

Applications are invited for a 9-month, tenure-track faculty position in food science at the assistant professor level in the Department of Human Nutrition. The successful candidate is expected to initiate a vigorous, externally funded research programme, maintain a high level of scholarly activity and train graduate students. In addition, he/she is expected to teach in the division's graduate and undergraduate programmes, and contribute to the development of campus-wide, interdisciplinary initiatives in life sciences, global health and/or sustainability. Service in department, college and campus committees and participation in national/international professional organisations are expected (5% effort). The successful candidate will demonstrate an ability to bridge the various disciplines within food science and human nutrition to address and promote human health. Multidisciplinary research opportunities are encouraged within the department and with the various centers on campus.

Qualifications
A PhD, MD or equivalent in nutrition, chemistry, zoology, biochemistry, physiology or related fields. Postdoctoral training preferred. Interest in furthering the research and teaching mission of the Division of Nutritional Sciences.

Academic rank and compensation
Assistant professor with academic year appointment and expectation of tenure. Position is designated as 50% effort in research, 50% in teaching/instruction.

Job 6. Principal nutrition scientist (nutrition company)

This international group is a global provider of high-quality technological ingredients and innovative solutions to the food, beverage and other industries. Reporting to the director, the job purpose is to identify, screen and recommend reachable opportunities in order to support the group nutrition strategy and speed up its position as a worldwide nutrition leader. His/her missions are to drive forecasting operations, to monitor nutritional development and to manage contacts with customers, opinion leaders, experts and scientific institutions in order to build networks useful for the group development. Internally, he/she will bring support, training and solutions to internal teams. Externally, he/she will represent the group in external audiences, participating in professional conventions and seminars.

Applicants should have:

- a Master's or PhD in nutrition, clinical research or related field
- 6–8 years of experience preferred in multiple areas of nutrition sciences
- excellent personal network through the scientific community and institutions
- multi-oriented project leader profile, at ease with any type of contacts
- leadership with a high sense of communication and representativeness.

The applicant must be ready to travel significantly to meet customers and visit group expertise centres.

CV 1 (chronological – academic)

Lauren BOURGET

25 rue de Campsaure, 6900 Lyon, France
Phone (mobile): +33 (0)6 63 01 33 93
Email: lauren.bourget@yahoo.fr
www.linkedin.com/pub/lauren-bourget/10/b72

RESEARCH INTERESTS & CAREER ASPIRATIONS

A highly motivated PhD graduate looking for a postdoctoral position with the long-term aim of achieving an academic research career. Proficient in a repertoire of cellular and molecular biology techniques. My principal research interests lie in drug discovery and development. The recent focus of my research has been on elucidating cell death mechanisms induced in prostate cancer.

EDUCATION

2008–2012 **PhD in Molecular and Cellular Biology,** Université de Lyon

Supervisor: Professor Jean Lazarus
Managed a research project in endocrinology & oncology in collaboration with urologists from Lyon-Ouest Hospital.
Project aim: Specified cell death mechanisms induced by ligand TRAIL in normal and cancer prostate in relation to hormonal status.

Strategy:
- Identified TRAIL-receptors implicated in cell death in rat prostate induced by castration;
- Combined treatments in vitro to improve cell death of cancer cells;
- Developed a mouse model of prostate cancer.

Publications and presentations:
Lazarus J, **Bourget L** (2011) "XXX XXX XXXX XXXXXXXXXXXXXXXXXXXXX".
Journal of Physiology (34), pp 345–352.
Bourget L, Lazarus J (2012) "XXX XXX XXXXX XXXXXXXXXXXXXXXXXXX"
submitted to *Prostate.*
Presentation to a French congress "XXXX XXXX XXXXX XXXXXXXXXXXX" in 2011.

2005–2007 **Master's in Biology,** Université de Lyon
Taught courses - Six months in Cellular Biology, Genetics, Physiology, Immunology and Neurobiology.
Internship - Four months in the laboratory INSERM U689 in Laennec University, Lyon. Participated in a project to identify the cellular proteins affected by herpes simplex virus type 1.

2003–2005 **Bachelor in Biology,** UCBL
Courses in Genetics, Cellular and Molecular Biology, Biochemistry, Physiology and Statistics.

RESEARCH EXPERIENCE

2007–2008 **Research Assistant,** CNRS/Institut Jacques Monod, Paris
Managed a research project for one year in a research team of five.
Project aims: Specified the role of an extracellular matrix protein on myoblaste differentiation.
Techniques used: Produced and purified a high quantity of recombinant proteins in mammal cells.
Tested these recombinant proteins in vitro on myoblaste adhesion, proliferation and differentiation.

SCIENTIFIC TECHNIQUES

Cellular biology	Routine culture cell lines (LNCaP, DU145, PC3, HEK193, C2C12), transient siRNA knockdowns, cellular tests (proliferation, adhesion and differentiation), apoptosis analysis (flow cytometry, DAPI, pro-apoptotic proteins detection and caspase activity dosage).
Molecular biology	Protein-RNA-DNA extraction, quantitative RT-PCR.
Protein biochemistry	Western-blot, co-immunoprecipitation, affinity chromatography, ELISA, notion in two-dimensional electrophoresis.
Animal experimentation	Prostate tumour generation in nude mice and follow-up the tumour growth.
Tissue Analysis	BOUIN fixation, paraffin-embedded tissue sections, TUNEL, immunohistochemistry.

PROJECT RESEARCH SKILLS

- Experimental protocols: designing, planning, budget-management and deadline respect;
- Effective presentations and reports on a regular basis; Problem solving;
- Training and management of two technicians.
- Language: French: Mother tongue; English: Fluent; Spanish: Basic conversation.
- Computing: Word, Excel, PowerPoint, EndNote, Adobe Photoshop, Statview, Optiquant.
- Other: Full driving licence.

TRAINING

March 2011 Patent rights sensitisation in INPI, Lyon
2009 Regulatory Training in Animal Experimentation (level I)

MEMBERSHIP AND INTERESTS

2011–2012 Member of "BioDocs-Lyon" association: researchers' network.
2010–2012 Member of "Alter-Conzo" association to promote responsible agriculture.
Sports Volley-Ball, squash, skiing, snowboarding, diving (PADI level I).

REFERENCES

Professor Jean Lazarus,
Researcher Professor,
Université de Lyon
Lyon, France
j.Lararusx@lyon.fr
+334.23.37.71.54

Dr Renée Pont
Senior scientist
CNRS/Institut Jacques Monod
Paris, France
r.Pontx@cnrs.fr
+33 1.23.45.91.32

CV 2 (targeted – technical research post)

Lauren BOURGET

25 rue de Campsaure, 6900 Lyon, France
Phone (mobile): +33 (0)6 63 01 33 93
Email: lauren.bourget@yahoo.fr
www.linkedin.com/pub/lauren-bourget/10/b72

RESEARCH INTERESTS & CAREER ASPIRATIONS

A highly motivated PhD graduate with over 5 years' professional experience in the laboratory. Proficient in a repertoire of cellular and molecular biology techniques, my career intention is to transfer my skills and core competencies into a drug discovery and development programme.

SCIENTIFIC TECHNIQUES

Cellular biology	Routine culture cell lines (LNCaP, DU145, PC3, HEK193, C2C12), transient siRNA knockdowns, cellular tests (proliferation, adhesion and differentiation), apoptosis analysis (flow cytometry, DAPI, pro-apoptotic proteins detection and caspase activity dosage), establishment of stable transfection in mammal cells, immunofluorescence.
Molecularbiology	Protein-RNA-DNA extraction, quantitative RT-PCR.
Protein biochemistry	Western-blot, co-immunoprecipitation, affinity chromatography, ELISA, notion in two-dimensional electrophoresis.
Animal experimentation	Prostate tumour generation in nude mice and follow-up the tumour growth.
Tissue Analysis	BOUIN fixation, parraffin-embedded tissue sections, TUNEL, immunohistochemistry.

EMPLOYMENT HISTORY

2007–2008 **Research Assistant,** CNRS/Institut Jacques Monod, Paris
Managed a research project for one year in a research team of five.
Project aims: Specified the role of an extracellular matrix protein on myoblaste differentiation.
Techniques used: Produced and purified a high quantity of recombinant proteins in mammal cells.
Tested these recombinant proteins in vitro on myoblaste adhesion, proliferation and differentiation.

EDUCATION

2008–2012 **PhD in Molecular and Cellular Biology,** Université de Lyon
Managed a research project in endocrinology & oncology areas in collaboration with urologists from Lyon-Ouest Hospital. Specified cell death mechanisms induced by ligand TRAIL in normal and cancer prostate in relation to hormonal status.

Identified TRAIL-receptors implicated in cell death in rat prostate induced by castration; Combined treatments in vitro to improve cell death of cancer cells; Developed a mouse model of prostate cancer.

2005–2007 **Master's in Biology,** Université de Lyon
Taught courses – Six months in Cellular Biology, Genetics, Physiology, Immunology and Neurobiology.
Internship – Four months in the laboratory INSERM U689 in Laennec University, Lyon. Participated in a project to identify the cellular proteins affected by herpes simplex virus type 1.

2003–2005 **Bachelor in Biology,** UCBL
Courses in Genetics, Cellular and Molecular Biology, Biochemistry, Physiology and Statistics.

PERSONAL AND TRANSFERABLE SKILLS

Teamworking and collaboration
Working as part of a research team comprising six members, liaising with other departmental colleagues and collaborating with external partners (university and industry);
Training and management of two technicians, taking responsibility for their work and offering advice.

Communication
Effective presentations at conferences and in the university department, writing reports on a

Problem solving
Able to think creatively and apply my knowledge to new problems. Discussing and brainstorming team members to work out solutions and find innovative ways to go forward.

PUBLICATIONS & PRESENTATIONS

Lazarus J, **Bourget L** (2011) "XXX XXX XXXX XXXXXXXXXXXXXXXXXXXXX". *Journal of Physiology* (34), pp 345–352.
Bourget L, Larazus J (2012) "XXX XXX XXXXX XXXXXXXXXXXXXXXXXXX" submitted to *Prostate*.
Presentation to a French congress "XXXX XXXX XXXXX XXXXXXXXXXXXX" in 2011.

TRAINING, MEMBERSHIP AND INTERESTS

March 2011	Patent rights sensitisation in INPI, Lyon.
2009	Regulatory Training in Animal Experimentation (level I).
2011–2012	Member of "BioDocs-Lyon" association: researchers' network.
2010–2012	Member of "Alter-Conzo" association to promote responsible agriculture.
Sports	Volley-Ball, squash, skiing, snowboarding, diving (PADI level I).

OTHERS SKILLS

Language:	French: Mother tongue; English: Good working knowledge; Spanish: Basic conversation.
Computing:	Word, Excel, PowerPoint, EndNote, Adobe Photoshop, Statview, Optiquant.
Other:	Full driving licence.

REFERENCES
Available on request.

CV 3 (targeted/skills based – science communication)

Lauren BOURGET

25 rue de Campsaure, 6900 Lyon, France
Phone (mobile): +33 (0)6 63 01 33 93
Email: lauren.bourget@yahoo.fr
www.linkedin.com/pub/lauren-bourget/10/b72; Fluent English

RELEVANT EXPERIENCE

Science communication and media

- Responsible for all media and communications whilst working with Alter-Conzo to promote responsible agriculture.
- Devised a handbook for members on media and press engagement in order to pass it on to future voluntary media officers.
- Familiar with the press avenues at the international level including Eurekalert and Alphagalileo.
- Organised a public event including contacting the press, local publicity campaign and press release.
- Presented talks at a departmental and national levels during conferences in English. Also presented posters at international conferences where I interacted and networked with other scientists.

Teamwork and collaboration

- Working as part of a research team comprising six members, liaising with other departmental colleagues and collaborating with external partners (university and industry);
- Liaised with other students based in Germany and UK in the collaborative network and had two lab visits during my PhD.
- Supervised and monitored the work of master's and degree students in the lab. Also, some tutorial work.

Project management

- Organising a major public event working as the voluntary media officer for Alter-Conzo required careful management to ensure the event was successful. This involved agreeing an initial team strategy, delegation of tasks, identifying local groups to target publicity, booking a venue, and liaising with the press and local officials.
- Following the event I wrote a report for the association to be distributed to other members in the network.
- As an active member of the BioDocs-Lyon" association: researchers' network I managed a schedule of events during the year to bring speakers into the institute to deliver afternoon lectures. With two other members of the group, I organised their travel and accommodation and took them for an evening meal.
- Project management of my PhD was essential due to its collaborative nature. It was necessary to prioritise tasks and to coordinate activities with other partners in the network.

EMPLOYMENT HISTORY

2007–2008 **Research Assistant,** CNRS/Institut Jacques Monod, Paris
- Managed a research project for one year in a research team of five.
- *Project aims*: Specified the role of an extracellular matrix protein on myoblaste differentiation.

EDUCATION

2008–2012 **PhD in Molecular and Cellular Biology,** Université de Lyon
Managed a research project in endocrinology & oncology areas in collaboration with urologists from Lyon-Ouest Hospital. Specified cell death mechanisms induced by ligand TRAIL in normal and cancer prostate in relation to hormonal status.

2005–2007 **Master's in Biology,** Université de Lyon
Taught courses - Six months in Cellular Biology, Genetics, Physiology, Immunology and Neurobiology.
Internship - Four months in the laboratory INSERM U689 in Laennec University, Lyon. Participated in a project to identify the cellular proteins affected by herpes simplex virus type 1.

2003–2005 **Bachelor in Biology,** UCBL
Courses in Genetics, Cellular and Molecular Biology, Biochemistry, Physiology and Statistics.

PUBLICATIONS & PRESENTATIONS

- Lazarus J, **Bourget L** (2011) "XXX XXX XXXX XXXXXXXXXXXXXXXXXXXXX". *Journal of Physiology* (34), pp 345–352.
- **Bourget L**, Larazus J (2012) "XXX XXX XXXXX XXXXXXXXXXXXXXXXXX" submitted to *Prostate*.
- Presentation to a French congress "XXXX XXXX XXXXX XXXXXXXXXXXX" in 2011.

TRAINING

March 2008	Patent rights sensitisation in INPI, Lyon
2005	Regulatory Training in Animal Experimentation (level I), UCBL
1997–1999	Training as Doctor of Medicine (the first year), Laennec University, Lyon

OTHERS SKILLS

Language:	French; Mother tongue; English: Fluent; Spanish: Basic conversation.
Computing:	Word, Excel, PowerPoint, EndNote, Adobe Photoshop, Statview, Optiquant.
Other:	Full driving licence.

MEMBERSHIP AND INTERESTS

2007–2008	Member of "BioDocs-Lyon" association: researchers' network.
2006–2008	Member of "Alter-Conzo" association to promote responsible agriculture.
Sports	Volley-Ball, squash, skiing, snowboarding, diving (PADI level I).

REFERENCES

Professor Jean Lazarus,
Researcher Professor,
Universite de Lyon,
Lyon
France
j.Lararusx@lyon.fr
+334.23.37.71.54

M. Sebastien Renoir
Alter-Conzo,
66 avenue des Bruyees
Lyon
France
alterconso@lyon.fr
+334.26.32.91.12

CV 4 (skills based – non-scientific role)

Lauren BOURGET

25 rue de Campsaure, 6900 Lyon, France
Phone (mobile): +33 (0)6 63 01 33 93
Email: lauren.bourget@yahoo.fr
www.linkedin.com/pub/lauren-bourget/10/b72; Fluent English

CAREER MOTIVATION

A highly motivated PhD graduate looking to train into a career as a chartered accountant. Excellent communication and teamworking skills coupled with well developed analytical and problem solving abilities, my aim is to pursue an intellectually stimulating and challenging career.

PERSONAL AND TRANSFERABLE SKILLS

Academic ability

- Qualified to PhD level, I have achieved a consistent academic record throughout my education;
- Able to understand and synthesise complex information and present it in writing;
- Able to learn new concepts, facts and methodologies during my PhD. This included reading research literature, learning new computer packages and laboratory techniques.

Teamworking and collaboration

- Working as part of a research team comprising six members, I liaised with other departmental colleagues and collaborating with external partners (university and industry);
- As an active member of the university postgraduate association I organised events with other students and researchers. For example, arranging outside professionals to speak in the department and monthly social events.
- Trained and managed two technicians, taking responsibility for their work and offering advice.

Commercial awareness

- Due to my interest in commercialising research, I took a training course in patent rights;
- Organising talks in the department required adherence to a budget for speaker travel expenses.

Communication

- Made oral presentations to the university department and to a congress;
- During my role as research assistant I wrote a final report of my findings which was presented to the other team members.

Curiosity and Problem solving

- Able to think creatively and apply my knowledge to new problems;
- Discuss and brainstorming team members to find solutions and innovative ways to go forward.

Flexibility and Adaptability

- Conducting new research during my PhD, I was faced with unpredictable results requiring critical thinking and a flexible approach;
- Time management meant working weekends and long hours according to experimental protocols especially during long-term studies;

EDUCATION

2008–2012 **PhD in Molecular and Cellular Biology,** Université de Lyon
Managed a research project in endocrinology & oncology areas in collaboration with urologists from Lyon-Ouest Hospital. Specified cell death mechanisms induced in normal and cancer prostate in relation to hormonal status.

2005–2007 **Master's in Biology,** Université de Lyon

2003–2005 **Bachelor in Biology,** UCBL

EMPLOYMENT HISTORY

2007–2008 **Research Assistant,** CNRS/Institut Jacques Monod, Paris
Managed a research project for one year in a research team of five.

TRAINING, MEMBERSHIP AND INTERESTS

March 2011	Patent rights sensitisation in INPI, Lyon.
2009	Regulatory Training in Animal Experimentation (level I).
2011–2012	Member of "BioDocs-Lyon" association: researchers' network.
2010–2012	Member of "Alter-Conso" association to promote responsible agriculture.
Sports	Volley-Ball, squash, skiing, snowboarding, diving (PADI level I).
Travelling	New-Caledonia, Europe, Morocco.

OTHERS SKILLS

Language:	French: Mother tongue; English: Good working knowledge; Spanish: Basic conversation.
Computing:	Word, Excel, PowerPoint, EndNote, Adobe Photoshop, Statview, Optiquant.
Other:	Full driving licence.

REFERENCES

Professor Jean Lazarus,
Researcher Professor,
Université de Lyon
Lyon
France
j.LeFrance@nice.fr
+334.23.37.71.54

M. Sebastien Renoir
Alter-Conso,
66 avenue des Bruyees
Lyon
France
alterconso@lyon.fr
+334.26.32.91.12

CV 5 (chronological – academic)

JAMES APPLEGARTH

1001 London Drive
Greenfield Heights
Chicago MI 45362
USA
Email: j.applegarthx@chicago.edu
Phone (mobile): +1 (0) 7566 630123
www.linkedin.com/jamesapplegarth/10/b22/456
Blog: www.bioscicareer.org

KEY CAPABILITIES

- Highly experienced postdoctoral associate focussing on research into specific inter-actions between humans and animals.
- Six peer-reviewed primary papers; one review paper; journal referee.
- Co-secured funding for international collaboration with research groups in four countries (in Europe and South America) and industrial stakeholders.
- Supervise two PhD students.

RESEARCH EXPERIENCE

2009– present   **Postdoctoral Associate,** Department of Zoology, University of Chicago
The effects of human-animal interactions on animal welfare and product quality.
Funded by an independent fellowship secured from the NSF in 2008, my research focuses on the specific interactions between animals and humans to determine the effect on the nutritional value and quality of the resulting food product. Working as part of an international research consortium, I am using biochemical and molecular techniques to determine specific nutritional content as well as surveying public opinion. The aim and impact of the research is to optimise animal welfare conditions and to influence the supply to food retail companies.

KEY PUBLICATIONS: See page 3 for full list

Primary paper: **Applegarth, J**., Smith, C.T and Mouiller, T.2011. XXXXXXXXXXXXXX
Review paper: **Applegarth, J**. 2010. XXXXXXXXXXXXXX

AWARDS and ACHIEVEMENTS

- Poster prize during Experimental Biology conference (2009)
- Best student (final year undergraduate), University of Nevada (2004)

SUPERVISION and TEACHING

- Currently supervise two graduate students. Analysed and discussed their work and explained scientific work as well as literature search to them.
- Acted as tutor in practicals throughout my PhD.

IMPACT ACTIVITIES:

- Worldwide media coverage as a result of press release following key findings in 2011.
- Presented PhD project on the radio and gave an interview on food standards and quality.

- Took part in a public event organised by the Institute. Explained ongoing projects to schoolchildren and parents, international guests and authorities.
- Public dialogue through regular blogs.

RESEARCH TECHNIQUES

- Biochemistry: Fatty acid chemistry, lipid extraction, thin layer chromatography, gas chromatography;
- Physiology: Respirometry, energy assimilation, nutritional physiology, gastrointestinal physiology, energy content calculation;
- Statistics: able to analyse large datasets, analyses of variance, generalized linear model, phylogenetic generalised least squares regression;
- Animal keeping: familiar with maintenance and needs of lab rodents and lagomorphs.

ADDITIONAL SCIENCE PROJECTS AND EXPERIENCE

- Undertook voluntary work for a food policy organisation (Food Ethics Council) which comprised analysing blood samples of farm animals (including cows, sheep, pigs). Participated in field work by collecting faecal samples of livestock, recorded fatty acid content of ruminants.
- Produced a comprehensive report of findings – available on request.
- Researched literature associated with US food policy issues and collated the review into a document which was used by the Food Ethics Council towards a national document submitted to government.

EDUCATION

2005–2009   **PhD Zoology,** Department of Animal Sciences, University of Boston,
Supervisor: Professor Julia Gonzalez
Using gas chromatography, muscle fatty acid composition was examined in 25 mammalian species to determine nutritional content resulting in one publication (see page 3).

2004–2005   **MSc in Nutrition,** School of Food and Nutrition, University of Nevada
Title of Diploma: Nutritional and fatty acid biosynthesis regulated by leptin gene expression PLUS2/DRED.

2001–2004   **BSc Zoology,** University of Nevada
Main subjects studied: animal physiology, parasitology, wildlife biology.

REFEREES

Associate Professor Dr. T. Moullier
Department of Zoology,
University of Chicago
Chicago, MI
Tel.: +1 (0) 7566 XXXXX
Fax: +1 (0) 7566 XXXXX
thomas.moullierx@nevada .edu

Professor Dr. Julia Gonzalez
Department of Animal Sciences,
University of Boston
Tel.: +1 (0) 2366 XXXXX
Fax: +1 (0) 2345 XXXXX
j.gonzalezx@boston.ac.es

CV 6 (targeted – industry)

JAMES APPLEGARTH

1001 London Drive
Greenfield Heights
Chicago MI 45362
USA
Phone (mobile): +1 (0) 7566 630123
Email: j.applegarthx@chicago.edu
www.linkedin.com/jamesapplegarth/10/b22/456

KEY CAPABILITIES

- PhD-qualified nutritionist with seven years' experience in the research field.
- Excellent communication skills, team-orientated internally and externally with business partners.
- Highly developed networking skills on an international platform.
- Supervision and leadership of students; identification of development needs.

RESEARCH AND MANAGEMENT EXPERIENCE (NUTRITION)

Nutrition expertise

- Well developed expertise in the field of nutrition spanning 7 years.
- Knowledgeable about potential new opportunities through research, contacts and collaborators (academic and industrial).
- Excellent research track record with six peer-reviewed papers and a review published during the course of my research.
- Up to date with international nutritional developments including food policy.
- Extensive experience in nutritional scientific and technical analysis.

Leadership and motivation

- Highly self-motivated and able to motivate and lead others in projects to achieve results within a specified time-frame.
- Working in a research team with external collaborators, I have co-ordinated the activities of the network through email and face-to-face visits.
- Supervised the work of students within the laboratory throughout the course of my career and have attended courses on leadership and team management.
- Work to high standards as evidenced in the awards achieved during my career.

Communication and teamwork

- Well developed communication skills. I make regular presentations internally and externally at international conferences to audiences of up to 250.
- Highly capable of initiating introductions and forming new collaborations through informal conversations and targeted networking.
- Interact with a wide range of interdisciplinary researchers and business partners as well as conducting public engagement activities all requiring versatile communication formats.
- Contributed to a review document used by the Food Ethics Council towards a national document submitted to government.

EMPLOYMENT HISTORY

2009– present **Postdoctoral Associate,** Department of Zoology, University of Chicago in collaboration with Foodworks Inc.
The effects of human-animal interactions on animal welfare and product quality.

- Funded by a fellowship secured from the NSF in 2008, the research focuses on specific interactions between animals and humans to determine the effect on the nutritional value and quality of the resulting food product.
- The research is a collaborative initiative with a local food supplier, Foodworks Inc. My role is to liaise between the university and all research partners to identify opportunities to improve product quality.

RESEARCH TECHNIQUES

- Biochemistry: Fatty acid chemistry, lipid extraction, thin layer chromatography, gas chromatography;
- Physiology: Respirometry, energy assimilation, nutritional physiology, gastrointestinal physiology, energy content calculation;
- Statistics: able to analyse large datasets, analyses of variance, generalized linear model, phylogenetic generalised least squares regression;
- Animal keeping: familiar with maintenance and needs of lab rodents and lagomorphs.

EDUCATION

2005–2009 **PhD Zoology,** Department of Animal Sciences, University of Boston, Using gas chromatography, muscle fatty acid composition was examined in 25 mammalian species to determine nutritional content resulting in one publication.

2004–2005 **MSc in Nutrition,** School of Food and Nutrition, University of Nevada
Title of Diploma: Nutritional and fatty acid biosynthesis regulated by leptin gene expression PLUS2/DRED

2001–2004 **BSc Zoology,** University of Nevada
Main subjects studied: animal physiology, parasitology, wildlife biology

REFEREES

Associate Professor Dr. T. Moullier
Department of Zoology,
University of Chicago
Chicago, MI
Tel.: +1 (0) 7566 XXXXX
Fax: +1 (0) 7566 XXXXX
thomas.moullierx@nevada .edu

Mr Francis Johnson
Director
Foodworks Inc.
Chicago, MI
Tel.: +1 (0) 7566 XXXXX
Fax: +1 (0) 7566 XXXXX
f.johnson@foodworks.com

- *Analysis A*: CVs 1, 2, 3 and 4 belong to Lauren Bourget who has just completed her PhD. They have been adapted and targeted to each of the corresponding jobs 1, 2, 3 and 4.
- *Analysis B*: CVs 5 and 6 belong to James Applegarth who is nearing the end of his first 3-year postdoctoral post. They have been restructured according to the specifications of jobs 5 and 6.

Analysis A (CVs 1, 2, 3 and 4)

Lauren Bourget has just completed her PhD in molecular and cell biology at the University of Lyon. During this time she has acquired many technical skills and expertise from her research on cell death mechanisms associated with prostate cancer. She has also developed personal skills such as planning, communication and working to deadlines. Alongside her PhD, Lauren has been involved in her university doctoral association (BioDocsLyon) and a local association for responsible agriculture (Alter-Conzo). Prior to undertaking her PhD, Lauren did a Master's degree followed by a research assistantship in Paris.

Lauren has built up a lot of experience and could take her career in many directions. Her PhD has given her in-depth knowledge of her research specialisation but her research skills could be transferred to a range of other research and technical roles. She has developed other skills and experience alongside her PhD which she could transfer into non-research or non-scientific roles, depending on where her interests lie (see Chapter 3). The following analyses of CVs 1–4 demonstrate the different ways in which Lauren has reorganised her CV to make it relevant to a variety of jobs.

CV 1 (chronological – academic)

CV 1 is targeted to job 1 (postdoctoral research scientist, research institute). Job 1 is a postdoctoral post and is described in detailed terms, which specify the techniques required by applicants as well as asking for a good publication record.

CV 1 includes a 'Career goal' at the top of page 1 which is linked to the job specification. It includes information on techniques and research interests relevant to the post. Lauren has highlighted her PhD by placing the 'Education' section first with a lengthy description of the research project, including its aims and corresponding list of publications (if this had been a longer list, Lauren could have stated here 'see Appendix for list of publications and presentations' and then appended them to the CV). Because this is an academic CV, the whole of page 1 is taken up with educational information and research experience.

Page 2 provides a list of research techniques which are important to the post with additional skills, training and membership information mentioned briefly in the second half of this page. Her referees are her supervisors from her PhD and research post, who will be able to vouch for her academic ability.

CV 2 (targeted – technical research post)

CV 2 is targeted to job 2 (scientific officer, research institute). Although this post, like job 1, is based in a research institute, job 2 is a supportive technical role.

Standard and specialised laboratory techniques and research experience are required of the applicants. There is no mention of published papers in the job requirements; instead personal skills are specified such as problem solving, team working and communication.

In CV2, the 'Career goal' is oriented towards this job as it describes the academic research as 'professional laboratory experience'. Lauren has positioned her list of scientific techniques on page 1 and her research post follows directly on from this, highlighting her employment experience. Academic experience and evidence of publications are not as important to this role as for job 1, but a PhD is 'highly desirable' and so appears at the end of page 1. Personal skills are more extensively described and are linked to those specified in the vacancy criteria. Lauren has chosen not to include her referees in this CV, but the same referees she used in CV1 would be appropriate.

CV 3 (targeted/skills based – science communication)

CV 3 is targeted to job 3 (communications manager, European science organisation). Job 3 is a communications post based in a scientific organisation. Although a PhD is not specified in the criteria, it proves an interest in life science research, which is one of the main stated requirements for the post. The key responsibilities of the post are not directly related to Lauren's research as with jobs 1 and 2. However, her voluntary experience working with her university doctoral association and the local agricultural association are highly applicable here.

CV 3 is highly targeted to this communications job. Lauren has chosen not to include a 'Career goal' in this CV. Instead, page 1 is taken up almost entirely with three substantial sections, which have been titled with the identical skills as stated in the job requirements. Using bullet points, Lauren has detailed all the tasks she has carried out in her voluntary activities which evidence her suitability for this role. Although her PhD information is on page 2, she mentions it in one of the bullet points and also links the skills and experience gained from her research to this post. The description of her PhD is given less space as its specifics will not be of interest to the employers, nor will her scientific techniques, which do not feature on CV3. Lauren has used her PhD supervisor as one referee and the other is the director of Alter-Conzo, who can vouch for her communications ability.

CV 4 (skills based – non-scientific role)

CV 4 is targeted to job 4 (trainee chartered accountant, international accountancy firm). The description and criteria described for job 4 make no mention of specific qualifications, but do state the need for candidates to have a 'good academic ability'. The job specifications are sectioned into five personal skills – flexibility, curiosity, team working, commercial awareness and great communication.

Although the application method is by online application form, CV4 demonstrates how you could set out your information to highlight your transferable skills. Page 1 is focused entirely on each of the skill-sets identified in the personal specifications. Bullet-pointed examples are used to demonstrate academic ability, oral and written communication skills, flexibility and all the other skills specified in the job description. Education and employment follow on page 2 documenting from where Lauren acquired her experience and transferable skills. She has used her supervisor as a referee to vouch for her academic ability and her manger at Alter-Conzo to give a different perspective of her experience.

Analysis B (CVs 5 and 6)

James Applegarth is nearing completion of an independent postdoctoral fellowship. Working at the University of Chicago, James' research focuses on human–animal interactions and the effect on nutritional quality of food products. Previously, he undertook a PhD in nutrition which started his interest in this research area. James is working within an international collaboration and has built up a list of publications, including writing a review. He supervises and tutors students and has won academic prizes.

James' academic career to date places him in a good position to take his academic career further. He has built up a good publication record and is establishing himself within his field. His collaborative research is giving him contacts and the opportunity to find a niche to pursue his research. His additional research responsibilities such as communications, supervision and teaching provide him with extra skills to widen his range of future career options (see career narratives 1–3 in Appendix 1 to read more about academic careers). CVs 5 and 6 demonstrate how James could tailor his CV to an academic post or to a senior post in industry.

CV 5 (chronological – academic)

CV 5 is targeted to job 5 (senior academic role, US university). Job 5 is an assistant professorship* leading to a permanent academic post in the Department of Human Nutrition. The requirements of the post are typical of any academic research position, e.g. able to initiate and lead a research programme, evidence of secured funding, excellent publication record, willing to teach undergraduate students and undertake departmental administrative duties.

In CV5 James has used a reverse chronological CV to match his highly relevant and specific research experience and skills to the job specification. He has listed four 'Key capabilities' at the start of his CV to highlight his research record more significantly to ensure he gets noticed during the selection process (academic posts can sometimes attract over 100 applicants). James has listed his two key publications on page 1 and has appended the full list to the CV. Although the précis of his research is quite short in his CV, he will have been required to send a 1–2-page research statement setting out the aims and direction of his research in more depth.

On page 2 of his CV, James has included additional information of interest to the nutrition department such as his research techniques and additional engagement with his subject beyond research, such as policy and business. His education is listed at the end of page 2. Since a PhD is mandatory to the post, it would have been a waste of valuable space to place this on page 1, since it would not differentiate him from the other candidates.

CV 6 (targeted - industry)

CV 6 is targeted to job 6 (principal nutrition scientist, nutrition company). Job 6 is a senior strategic role in industry requiring the post holder to forecast and manage the direction of the nutritional operations. The vocabulary used in the

---

* Note that academic posts vary in their titles according to different countries – see Chapter 4, page 40 for a fuller explanation).

advertisement is highly dynamic, e.g. 'recommend reachable opportunities', 'drive forecasting operations' and 'manage contacts with opinion leaders'. The responsibilities and working environment are not as familiar to James as those associated with academia but if he is interested in working in industry, he can tailor his CV and vocabulary to the job specification.

Although James can match himself to many of the competencies required for the post, his lack of directly relevant experience in industry will place him in a weaker position than for a senior academic post. Therefore, in order to be selected for interview, James has reorganised and restructured his research experience in CV6 in line with the specifications of this role. Not only does this demonstrate the transferability and relevance of his experience but by using the language of business, he is showing his enthusiasm for moving out of academia into industry. The 'Key capabilities' section highlights his experience, so that it matches those specified in the job criteria. He has sectioned page 1 into the three key skills required for the post and bullet-pointed examples using dynamic vocabulary.

It is only on page 2 that James reveals himself to be an academic researcher and to show from where he has gained his experience. By this time, he is hoping to have portrayed a positive image of himself to the employer, who may be surprised about his background. Imagine if his 'Employment' section had been positioned first on page 1. The employer may have dismissed James out of hand because of his lack of industry experience before he had had the chance to 'sell' his marketable qualities, especially if there were many applicants for the job. The referees used in CV6 differ from those in CV5, with the substitution of one of the academic referees for one of James' business collaborators, who will be able to vouch for his commercially relevant abilities.

Analysis summary

These CVs demonstrate that you always need to alter your CV and the running order of its content in order to target it to the criteria, as specified in job advertisements. If the employer does not immediately see evidence of your suitability to their post, your CV may be overlooked. In order to get their attention and sell yourself, you need a strategic document. This is likely to produce more successful results than a generic academic CV, which may have served you well up until now.

Refer to Chapter 6 for guidelines on writing a CV and how to match it to particular jobs and their specifications. Advice on how to write a covering letter is also included in Chapter 6 (page 79) and an example is illustrated on the following page (165).

Covering letter

This example is based on CV3 which is being used to apply for job 3.

25 rue de Campsaure
6900 Lyon
France

date

Name of addressee (*if you have one*)
Company name and address

Communications Manager   (*include reference number if there is one*)

Dear sir/madam (*or named person*),

I am applying for the post of communications manager which I saw advertised on your website. Having recently completed my PhD in cell biology, I am very keen to pursue a career in science communication. I enclose my CV which provides evidence of my suitability for this post.

During the course of my PhD, I have been working as a media officer on a voluntary basis, in a local agricultural company. During this time I have gained a range of communications experience including media work, writing press releases and devising a handbook for media officers. I volunteered to take on this role due to my interest in writing science for more general audiences, as well as being motivated by working in a team with a common purpose. In addition, to media experience, I have also organised and delivered talks to the public and members of my department.

This role requires excellent organisational and project management skills which I have been developing through my research and voluntary work. As an active member of the university doctoral association, I have helped to organise a schedule of events during the year which involved inviting speakers and arranging their travel and accommodation, working within a budget. During my research, I collaborated with international researchers and helped to co-ordinate the network. This required regularly communicating with them by email or phone and making occasional visits to their labs (and vice versa).

The European Science Organisation is highly respected for its communications and other research-associated work and I would appreciate the opportunity of working with you in this exciting role. I hope you will consider me seriously for the post and look forward to hearing from you.

Yours faithfully, (if sir/madam)
Yours sincerely, (if named person)

Your name (PhD)

Support and resources

Many of us ask for advice informally from colleagues, friends and family looking for some sensible suggestions to help weigh up our career options in order to make decisions. This can be useful to clarify our thoughts and ideas, especially if the person is working in the career in which we are seeking advice. For example, your supervisor and other academics can advise you about research careers. You may have access to professionals working in other careers through your networks. Some of your friends or former colleagues could be working in fields of interest to you. However, for more general careers advice you may sometimes receive conflicting opinions which can cause more confusion than clarification. Therefore, much the same as with peer review in research, you need to look for reliable information, which derives from well-qualified, recognised experts in the field or specialist organisations.

The good news for research bioscientists is that, in the last decade or so, specialist support aimed specifically at doctoral students and postdoctoral research-ers has been building and expanding. National organisations and associations provide a wealth of resources, information and events aimed at assisting you with your career. You can usually access resources on their websites and you may be able to attend their events. In addition to the support you receive from your institution, you can also access specialist resources and support from funding councils, research organisations, learned societies, professional bodies and private consultancies. These range from academic-related training such as grant writing, publishing in journals and effective presentations through to leadership, entrepreneurship and management workshops, media training, career planning and CV writing. In addition, many scientific conferences run a supporting programme of career activities or organise networking events for early-career researcher delegates. There are even specialist conferences and job fairs aimed at scientists such as PhD Talent and Naturejobs Careers Expo. See www.biosciencecareers.org/p/education-policy-careers-meetings.html for regular updates on careers, education and policy meetings.

The career narratives in Appendix 1 provide profiles of people working in a range of professions with an additional commentary identifying factors which have influenced their career decisions and motivations.

Career Planning for Research Bioscientists, First Edition. Sarah Blackford.
© 2013 Sarah Blackford. Published 2013 by Blackwell Publishing Ltd.

Doctoral and postdoctoral support organisations

Country	Association/organisation	Web address / email contact
National organisations		
UK	Vitae	www.vitae.ac.uk
France	ABG/L'intelli'agence	www.intelliagence.fr
Germany	Kisswin	www.kisswin.de
Postdoc/doctoral associations		
USA	National Postdoctoral Association (NPA)	www.nationalpostdoc.org/
UK	UK Research Staff Association (UKRSA)	www.vitae.ac.uk/ researchers/205761/UK-Research-Staff-Association.html
France	ANDes	www.andes.asso.fr/
France	CJC Jeunes Chercheurs	http://cjc.jeunes-chercheurs.org/ presentation/index.en.php
Ireland	IRSA	www.irsa.ie
Norway	UIT Stipindat	www.forskerforbundet.no/
Finland		riku.matilainen@ tieteentekijoidenliitto.fi
The Netherlands	Postdoc Career Development Initiative (PCDI)	www.pcdi.nl/
Europe PGRs	Eurodoc	www.eurodoc.net/
Canada	Canadian Association of Postdocs (CAP)	http://sites.google.com/site/ canadapostdoc/Home
Brazil	SBPC	www.sbpcnet.org.br/site/home/
South Africa	National Research Foundation (NRF)	www.nrf.ac.za/
Australia	University of Western Australia Researchers' Association	www.research.uwa.edu.au/staff/ researchers-association
International	ICoRSA	www.icorsa.org
International	World Association of Young Scientists (WAYS)	http://ways.org/en
Science organisations		
Europe	EMBL	www.embl.it/research/
Europe	EMBO	www.embo.org/programmes/ embo-career-development.html
International	Learned societies	www.biosciencecareers.org/p/ learned-societies.html

Associations and organisations	
Association for Women in Science (AWIS) www.awis.org	AWIS is committed to helping women in STEM at every stage of their career achieve their greatest potential
Athena Forum www.athenaforum.org.uk	Provides a strategic oversight of developments that seek to advance the career progression and representation of women in STEMM in UK higher education
Daphne Jackson Trust www.daphnejackson.org	Returners scheme helping those with a background in STEM to return after a career break
Dorothy Hodgkin Fellowship Scheme http://royalsociety.org/grants/schemes/dorothy-hodgkin	This scheme is for early career scientists in the UK, who require a flexible working pattern due to personal circumstances such as parenting or caring responsibilities
L'Oreal Women in Science fellowship scheme www.womeninscience.co.uk	Aims to improve the position of women in science by recognising outstanding women researchers who have contributed to scientific progress
MentorNet www.mentornet.net	e-mentoring for diversity in engineering and science
MentorSET www.mentorset.org.uk	A UK mentoring scheme for women in STEM
UK Resource Centre (UKRC) www.theukrc.org/women	Works with a wide variety of organisations to offer training, consultancy and other support to female staff, management, members and students
Women into Science and Engineering (WISE) www.theukrc.org/get-involved/wise	Works with industry and education to inspire girls and attract them into STEM studies and careers
Articles, books and blogs	**Description**
Babcock L, Laschever S. (2008) *Women Don't Ask. Negotiation and the gender divide*. Bantam: New York.	This book shows women how to reframe their interactions and more accurately evaluate their opportunities
Athene Donald's blog http://occamstypewriter.org/athenedonald	Dame Athene Donald FRS is professor of physics at Cambridge University. She regularly blogs about women in science issues
Jarrett V, Tchen T. (2011) Helping women meet their economic potential. *Washington Post Opinions*. http://wapo.st/xxAK9t	An opinion article about policy initiatives to improve economic opportunities for women in the USA
Shinton S. (2009) Career break, not broken. *Analytical and Bioanalytical Chemistry* **394**, 1509–11. www.springerlink.com/content/j8r0176259u16181/fulltext.pdf	Article centring on career breaks and strategies to make a successful return to work

Science jobs and funding

Academic

Jobs in Europe/mobility	http://ec.europa.eu/euraxess/
Jobs and information for grads and postdocs	http://jobs.phds.org/
Postdoc jobs	http://postdoc.com/
Academic teaching posts	www.careeroverview.com/teachers-postsecondary-careers.html
Euroscience jobs	www.eurosciencejobs.com/jobs/biology
European Union Gateway	http://europa.eu/quick-links/job-seekers/
Academic jobs EU	www.academicjobseu.com/
Academic and academic-related jobs UK	www.jobs.ac.uk
Jobs in Germany, Austria, Switzerland	www.academics.com/
Jobs in Belgium and international	http://doctorat.be/
An academic career	www.academiccareer.manchester.ac.uk/
Resources for graduates and postdocs (US)	www.indiana.edu/~halllab/grad_resources.html

General

Nature	www.naturejobs.com
New Scientist	www.newscientistjobs.com/
Science	http://sciencecareers.sciencemag.org/
The Scientist	http://the-scientist.com/
Careers away from the bench	http://bit.ly/Ar79wH
Career options for PhDs	http://bit.ly/w6maot
Entering industry (doctoral students)	http://bit.ly/t3uwr7
Entering industry (postdoctorals)	http://bit.ly/topOaE
LinkedIn groups	www.LinkedIn.com– *search bio-companies*

Subject-specific sectors

Biotechnology/clinical jobs	www.pharmiweb.com
Biotechnology jobs	www.bioindustry.org/home/ – *search membership list*
Biomedical jobs (UK only)	www.careerscene.com/
Environmental jobs	www.stopdodo.com/
Plant science jobs	www.aspb.org www.gatsbyplants.leeds.ac.uk/careers/jobs.html
Pharma jobs	www.wileypharmaceuticaljobs.com/
Contract research organisations	www.biores.org/dir/Companies/Contract_Research_Organizations/
Pharmacology societies	www.meduni-graz.at/pharma/pharma-www/main-soc.htm

Also sign up to scientific recruitment consultancies and agencies in your country. View company websites.

Funding

EMBO – funding, training and mobility	http://mobility.embo.org/
Funding sources	http://sciencecareers.sciencemag.org/funding

Human Frontier Science Program	www.hfsp.org/
Marie Curie funding	http://ec.europa.eu/research/ mariecurieactions/
National Institute of Health (NIH) funding	http://grants.nih.gov/grants/oer.htm
National Science Foundation	www.nsf.gov/funding/
Policy and funding information	http://ResearchResearch.com

Science-related jobs

Specialist and technical administration

European Commission Civil Service	http://ec.europa.eu/civil_service/job/index_ en.htm
EU careers	www.eu-careers.eu/
Chartered patent attorneys	www.insidecareers.co.uk/patent
European Patent Office	www.epo.org/about-us/jobs.html
US Patent and Trademark office	http://usptocareers.gov/
Regulatory Affairs Professional Society	www.raps.org/

Clinical trials

Recruitech	www.clin-ops.com
Jobs4dd	www.jobs4dd.com
CK Science	www.ckscience.co.uk

Communication

PSCI-COM (discussion list)	www.jiscmail.ac.uk/lists/psci-com.html
Australian Science communicators	www.asc.asn.au/about-using-the-asc-email-lists/
European Medical Writers Association	www.emwa.org
Medical communication information	www.medcommsnetworking.co.uk/startingout/
American Medical Writers Association	www.amwa.org
National Association of Science Writers	www.nasw.org
Association of British Science Writers	www.absw.org
Society of Young Publishers	www.thesyp.org.uk/
Book careers	http://bookcareers.com/
Eurekalert	www.eurekalert.org
Alphagalileo	www.alphagalileo.org
Euroscience	www.euroscience.org
Ecsite	www.ecsite.eu/
European Science Foundation	www.esf.org/
European Science Education Gateway	www.xplora.org/ww/en/pub/xplora/index.htm
AAAS	www.aaas.org/
Association of Science – Technology Centres	www.astc.org
British Science Association	www.britishscienceassociation.org
Café Scientifique	www.cafescientifique.org/europe-links.htm

Learned societies (membership, grants and careers)

For a comprehensive up-to-date list go to: www.biosciencecareers.org/p/learned-societies.html.

Society for Experimental Biology	www.sebiology.org
Federation of European Biochemical Societies	www.febs.org
Biochemical Society	www.biochemistry.org
British Ecological Society	www.britishecologicalsociety.org
Society of Biology	www.societyofbiology.org
Developmental Biology	www.sfbd.fr/
American Society of Plant Biologists	www.aspb.org
Societé Francais Ecologie	www.sfecologie.org/
British Pharmacological Society (BPS)	www.bps.ac.uk/
Society for General Microbiology	www.sgm.ac.uk
Society for Endocrinology	www.endocrinology.org/
American Physiological Society (APS)	www.the-aps.org/
The Physiological Society	www.physoc.org
Institute of Biomedical Science	www.ibms.org/

Applications and personal development

Assessment centres	www.vitae.ac.uk/researchers/1384/Assessment%20 Centres.html
Developing your career	www.vitae.ac.uk/researchers/1303/Developing-your-career.html
Effective CVs	www.vitae.ac.uk/researchers/1372/Creating-Effective-CVs.html
Interview tests and exercises	www.prospects.ac.uk/assessment_centres.htm
Successful applications	www.vitae.ac.uk/researchers/1376/Job%20applications.html
Successful interviews	www.vitae.ac.uk/researchers/1380/Successful%20 interviews.html
YES Biotechnology	www.biotechnologyyes.co.uk/

Social media

Twitter for scientists	http://t.co/St1BwuST
Social media: a guide for researchers	www.rin.ac.uk/our-work/communicating-and-disseminating-research/social-media-guide-researchers
LinkedIn	www.linkedin.com
	– join groups relevant to your field of interest

Blog

Bioscience careers blog	www.biosciencecareers.org
Bio careers	www.biocareers.com
Bioindustry	www.bioinsights.com/
Careers after bioscience	http://biosciencecareers.wordpress.com/
Postdoc blogs	www.postdocsforum.com
	www.vitae.ac.uk/researchers/156431/Research-staff-blog.html
Nature Network blogs	http://network.nature.com/blogs
Life in the lab	http://blog.labguru.com/
Laboratory webzine	http://jennyrohn.com/lablit

Career-related Twitter accounts

@BiosciCareer
@Biocareers
@Eurodoc
@vitae_news
@PostdocsForum
@postgradtoolbox

Further reading

General careers

Ali L, Graham B. (2000) *Moving on in Your Career – a guide for academic researchers and postgraduates*. Cambridge: Routledge/Falmer.

Newhouse M. (1993) *Outside the Ivory Tower – a guide for academics considering career alternatives*. Cambridge, MA: OCS/Harvard University.

Robbins-Roth C. (2006) *Alternative Careers in Science. Leaving the ivory tower*, 2nd edn. Burlington, MA: Elsevier.

Academic careers

Blaxter L, Hughes C, Tight M. (1998) *The Academic Career Handbook*. Buckingham: Open University Press.

Bloomfield VA, El-Fakanhany, EE. (2008) *The Chicago Guide to your Career in Science: A toolkit for students and postdocs*. Chicago: University of Chicago Press.

Boden R, Epstein D, Kenway J. (2006) *Academic's Support Kit*. London: Sage.

Davidson C, Ambrose S. (1994) *The New Professor's Handbook*. Chichester: Wiley Publishing.

Dee P. (2006) *Building a Successful Academic Career in Scientific Research*. Cambridge: Cambridge University Press.

Delamont S, Atkinson P. (2005) *Successful Research Careers*. Milton Keynes: Open University Press.

Goldsmith J, Komlos J, Gold P. (2001) *Your Academic Career*. Chicago: Chicago University Press.

Johnson AM. (2009) *Charting a Course for a Successful Research Career*, 2nd edn. Amsterdam: Elsevier.

Ketteridge S, Marshall S, Fry H. (2002) *The Effective Academic*. London: Kogan Page.

Career planning

Bolles RN. (2012) *What Color Is Your Parachute? 2012: a practical manual for job-hunters and career-changers: 40th anniversary edition*. Berkeley, CA: Ten Speed Press.

Hawkins P. (1999) *The Art of Building Windmills. Career tactics for the 21st century*. Liverpool: GIEU.

Krumboltz, JD, Lewin Al S. (2004) *Luck is no Accident. Making the most of happenstance in your life*. CA: Impact Publishers.

Personal and professional development

Adams D, Sparrow J. (2007) *Enterprise for Life Scientists*. Bloxham, UK: Scion Publishing Ltd.

Howard Hughes Medical Institute. (2006) *Making the Right Moves – a practical guide to scientific management for post docs and new faculty*. New York: Fireside, Simon and Schuster.

Applications and interviews

AGCAS. (2007) *Going for interviews*. Special Interest booklet. Graduate Prospects. Available from: www.prospects.ac.uk.

Corfield R. (1999) *Preparing Your Own CV*. London: Kogan Page.

Eggert M. (1998) *The Perfect Interview*. London: Random House Business Books.

Jackson T. (2004) *The Perfect CV*. London: Piatkus Books.

Miller R. (1998) *Promoting Yourself at Interview*. London: Trotman.

Shapiro M, Straw S. (1999) *Tackling Tough Interview Questions in a Week*. London: Hodder and Stoughton/Institute of Management.

Shavick A. (2005) *Management Level Psychometric and Assessment Tests*. Oxford: How To Books.
Tolley H, Wood R. (2010) *How to Succeed at an Assessment Centre*. London: Kogan Page.
Williams L. (2005) *The Ultimate Interview Book*. London: Kogan Page.
Yate MJ. (2003) *The Ultimate CV Book*. London: Kogan Page.
Yate MJ. (2008) *Great Answers to Tough Interview Questions*. London: Kogan Page.

Communication

Cargill M, O'Connor PJ. (2009) *Writing Scientific Research Articles*. Oxford: Wiley-Blackwell. Supporting website: www.writeresearch.com.au.
Day RA, Gastel B. (2006) *How to Write and Publish a Scientific Paper*, 6th edn. Westport, CT: Greenwood Press.
Kitchen R, Fuller D. (2005) *The Academic's Guide to Publishing*. London: Sage.
McConnon S. (2005) *Presentation with Power*. Oxford: How To Books.
Shephard K. (2005) *Presenting at Conferences, Seminars and Meetings*. London: Sage.
Thody A. (2006) *Writing and Presenting Research*. London: Sage.

Peer review

Hames I. (2007) *Peer Review and Manuscript Management in Scientific Journals: guidelines for good practice*. Oxford: Wiley-Blackwell.
Wager E, Godlee F, Jefferson T. (2002) *How to Survive Peer Review*. London: BMJ Books.

Index

Career Planning for Research Bioscientists, First Edition. Sarah Blackford.
© 2013 Sarah Blackford. Published 2013 by Blackwell Publishing Ltd.